As If By Design

The eureka moment is a myth. It is an altogether naïve and fanciful account of human progress. Innovations emerge from a much less mysterious combination of historical, circumstantial, and accidental influences. This book explores the origin and evolution of several important behavioral innovations including the high five, the Heimlich maneuver, the butterfly stroke, the moonwalk, and the Iowa caucus. Such creations' striking suitability to the situation and the moment appear ingeniously designed with foresight. However, more often than not, they actually arise "as if by design." Based on investigations into the histories of a wide range of innovations, Edward A. Wasserman reveals the nature of behavioral creativity. What surfaces is a fascinating web of causation involving three main factors: context, consequence, and coincidence. Focusing on the process rather than the product of innovation elevates behavior to the very center of the creative human endeavor.

EDWARD A. WASSERMAN is Professor of Psychology at the University of Iowa, USA, and has held visiting appointments in England, Russia, Japan, and France. He has published widely and received several prestigious awards in the areas of learning, memory, and cognition in people and animals.

As If By Design

How Creative Behaviors Really Evolve

EDWARD A. WASSERMAN
University of Iowa

CAMBRIDGE
UNIVERSITY PRESS

University Printing House, Cambridge CB2 8BS, United Kingdom

One Liberty Plaza, 20th Floor, New York, NY 10006, USA

477 Williamstown Road, Port Melbourne, VIC 3207, Australia

314–321, 3rd Floor, Plot 3, Splendor Forum, Jasola District Centre, New Delhi – 110025, India

103 Penang Road, #05–06/07, Visioncrest Commercial, Singapore 238467

Cambridge University Press is part of the University of Cambridge.

It furthers the University's mission by disseminating knowledge in the pursuit of education, learning, and research at the highest international levels of excellence.

www.cambridge.org
Information on this title: www.cambridge.org/9781108477765
DOI: 10.1017/9781108774895

© Edward A. Wasserman 2021

This publication is in copyright. Subject to statutory exception and to the provisions of relevant collective licensing agreements, no reproduction of any part may take place without the written permission of Cambridge University Press.

First published 2021

Printed in the United Kingdom by TJ Books Limited, Padstow Cornwall

A catalogue record for this publication is available from the British Library.

Library of Congress Cataloging-in-Publication Data
Names: Wasserman, Edward A., author.
Title: As if by design : how creative behaviors really evolve / Edward A. Wasserman.
Description: Cambridge, United Kingdom ; New York, NY : Cambridge University Press, 2021. | Includes bibliographical references and index.
Identifiers: LCCN 2020056639 (print) | LCCN 2020056640 (ebook) | ISBN 9781108477765 (hardback) | ISBN 9781108745109 (paperback) | ISBN 9781108774895 (epub)
Subjects: LCSH: Creative ability. | Creative thinking. | Inventions–Psychological aspects. | Technological innovations–Psychological aspects.
Classification: LCC BF408 .W29 2021 (print) | LCC BF408 (ebook) | DDC 153.3/5–dc23
LC record available at https://lccn.loc.gov/2020056639
LC ebook record available at https://lccn.loc.gov/2020056640s

ISBN 978-1-108-47776-5 Hardback
ISBN 978-1-108-74510-9 Paperback

Cambridge University Press has no responsibility for the persistence or accuracy of URLs for external or third-party internet websites referred to in this publication and does not guarantee that any content on such websites is, or will remain, accurate or appropriate.

To Leyre, as we just make it up as we go along

Contents

List of Figures		*page* x
Acknowledgments		xii
	SECTION 1 INTRODUCTION	1
	Prelude	3
	Setting the Stage	5
	Previewing the Vignettes	13
	SECTION 2 THE VIGNETTES	17
	PART I SPORTS	19
1	Dick Fosbury's High Jump Was No Flop!	21
2	Shedding Fresh Light on the History of the Butterfly Stroke	28
3	The Monkey Crouch: Jockeying for Position	39
4	Riding Acey-Deucy	49
5	The High Five: One Singular Sensation	55
	PART II MEDICINE	63
6	The Apgar Score: "Millions Have Been Saved"	65

7	The Ponseti Method: Effective Treatment for Clubfoot Only 2,400 Years in the Making	77
8	The Heimlich Maneuver	88
9	Eating to Live: The Lifesaving Contribution of Stanley Dudrick	97
10	What's in a (Drug) Name?	106
11	Self-Medication by People and Animals	112
12	Personalized Medicine: The End of Trial-and-Error Treatment?	121

PART III HYGIENE 131

13	Florence Nightingale: Advancing Hygiene through Data Visualization	133
14	Taking Mental Floss to Dental Floss	152
15	A Very Close Shave	162

PART IV ARTS, ENTERTAINMENT, AND CULTURE 171

16	Ansel Adams: Art for Art's Sake?	173
17	Basil Twist: "Genius" Puppeteer	192
18	Moonwalking: And More Mundane Modes of Moving	199
19	Play on Words	212

20 Cuatro Festivales Españoles   223

21 Tchaikovsky: Puzzles of the *Pathétique*   241

22 The Evolution of the Violin: Survival of the Fittest or the
    Fondest Fiddle?   250

    PART V   IS THIS HEAVEN? NO, IT'S IOWA!   257

23 The Rise and the Demise of the Iowa Caucus   259

24 Iowa's Blackout Plates: Artistic License Hits the Road   270

25 If I Write It, They Will Build It   276

    SECTION 3   PUTTING IT TOGETHER   285

26 Context, Consequence, and Coincidence   287

27 Are We Just Making It Up as We Go Along?   309

    Index   312

Figures

1.1	Dick Fosbury using his signature "Fosbury Flop."	*page* 22
2.1	Iowa Fieldhouse pool in the late 1920s.	32
2.2	Iowa coach David A. Armbruster in the 1940s.	33
2.3	Birthplace of the Butterfly. Banner hanging above the Olympic pool on The University of Iowa campus.	35
4.1	Riding with the acey-deucy positioning of leathers and stirrups.	51
7.1	Dr. Jose Morcuende applying a plaster cast to the right foot of an infant.	82
7.2	Dr. Ignacio Ponseti manually demonstrating the articulation of bones in the foot.	83
7.3	Dr. Ignacio Ponseti viewing an X-ray of a patient's feet.	86
9.1	Dr. Stanley J. Dudrick with one of his beagles, Stinky.	100
14.1	Monkey Chonpe-69-85-94 flossing her teeth with her own hair.	157
15.1	Older man from the Solomon Islands shaving with a clam shell while holding a mirror (1964).	166
16.1	Ansel Adams in 1953.	176
17.1	Basil Twist's team of backstage puppeteers in wetsuits rehearsing "Symphonie Fantastique."	196
18.1	Astronaut Buzz Aldrin walks on the surface of the moon near the leg of the lunar module Eagle during the Apollo 11 mission. Mission commander Neil Armstrong took this photograph with a 70 mm lunar surface camera.	200

18.2	The Partial Gravity Simulator, or POGO. The POGO supports 5/6th of the astronaut's weight in order to simulate walking on the lunar surface.	202
21.1	First passage (top two lines): the two lines of music that Tchaikovsky scored for the first and second violins at the opening Exposition of the final movement. Played separately for each group of violins, these notes make little sense. First passage with melody in gray (middle two lines): The emergent melody is highlighted in gray. These are the notes to the melody that we actually hear, even though no single instrument plays them. Second passage (bottom two lines): In the Recapitulation toward the end of the final movement, Tchaikovsky repeats the theme, but he assigns the first violins the melody shown above in gray and the second violins the harmony created from the remaining notes. There is no auditory trickery now.	243
24.1	Iowa's Dordt University license plate was originally issued in 2011 and was slightly changed in 2019 to reflect the school's name change (top). Iowa's new blackout license plate was issued in July of 2019 (bottom).	271
25.1	Author W. P. (Bill) Kinsella.	277

Acknowledgments

One of the origins of the word *acknowledgment* is an Old English word meaning "to understand" or "to come to recognize." Because this book took so long to produce, its own origins go back many years. Thinking back, I have come to understand and appreciate the many people who have contributed to this slowly developing project. I'll duly acknowledge them here.

It all began with a paper I wrote with my colleague Mark Blumberg in 1995, which concerned animal intelligence and William Paley's argument from design. In it, we critiqued the existence of a conscious designer inside the head of animals and people – the mind – which was responsible for adaptive and creative behavior. This mental designer bore a striking logical resemblance to another designer – a deity – which was responsible for the very creation of animals and people. Mark and I pursued this critique in other articles through 2010.

In 2009, a highly publicized *Science* paper reported the measurable benefits provided by the crouched position of a jockey on a racehorse. My research collaborator Leyre Castro urged me to continue my investigations into the origin of this behavioral innovation and several others, also believing that storytelling might be an especially effective way to extend the reach of scientific accounts of creativity and innovation.

Over the next several years, I published a few popular and scholastic pieces applying this rhetorical technique, all the while accumulating as many stories as I could that provided clues to the origins of creative behaviors. One key tip came from my cousin Ron, who was a horseracing aficionado. Others came from current events as well as historical markings of the birth or death of important innovators.

Several individuals helped in ways both big and small with the individual vignettes I've included in the book. I list them according to the vignette with which they helped.

Dick Fosbury's High Jump Was No Flop! Mark Blumberg, Tim Trower.
Shedding Fresh Light on the History of the Butterfly Stroke. Marc Long, Steve Roe, Bob Barney, Dave Barney, Paul Muhly.
The Monkey Crouch: Jockeying for Position. Mark Blumberg.
Riding Acey-Deucy. Ron Wasserman, Jason Neff.
The High Five: One Singular Sensation. Jon Mooallem.
The Ponseti Method: Effective Treatment for Clubfoot Only 2,400 Years in the Making. Jose Morcuende, Jodie Plumert.
Taking Mental Floss to Dental Floss. Jean-Baptiste Leca.
Basil Twist: 'Genius' Puppeteer. Basil Twist.
Moonwalking: And More Mundane Modes of Moving. Karen Adolph.
Cuatro Festivales Españoles. Leyre Castro, Verna Kale.
Tchaikovsky: Puzzles of the Pathétique. Christopher Russell, William LaRue Jones, Leyre Castro, David Wasserman, Tareq Abuissa.
The Evolution of the Violin: Survival of the Fittest or the Fondest Fiddle? Nick Makris, Dan Chitwood, David Schoenbaum.
Iowa's Blackout Plates: Artistic License Hits the Road. Paul Cornelius.

Still other people significantly contributed to my thinking and writing: Jonnie Hughes, Jack Marr, Ralph Miller, Michael Sauder, and Eric Scerri. I thank all of them and the vignette contributors for their help in making this book possible.

I also thank the College of Liberal Arts and Sciences for providing me with a Career Development Award during the 2018–2019 academic year to promote my progress on this endeavor.

The cover photograph depicts the bronze relief sculpture crafted by Larry Nowlan portraying Jack Sieg swimming an early version of the Butterfly Stroke. Cover photographer: Edward Wasserman.

Finally, I thank the American Psychological Association for permitting me to reproduce with minor editing privileges portions of this paper:

Wasserman, E. A. and Cullen, P. (2016). Evolution of the Violin: The Law of Effect in Action. *Journal of Experimental Psychology: Animal Learning and Cognition*, 42, 116–122.

SECTION I Introduction

Prelude

As you watch Michael Phelps swim the butterfly stroke, you marvel at the amazing coordination of his muscular arms, legs, and torso. These powerful motions seem to have been perfectly planned to outpace his rivals in what is the most demanding of all swimming strokes. What you may not appreciate is that the butterfly stroke was never planned. Nor, for decades following its debut, was it even an approved stroke in the Olympic Games.

In this book, I'll explore the origin and evolution of the butterfly stroke and a variety of other behavioral innovations which, because of their striking suitability to the situation, appear to have been ingeniously and foresightfully designed. More often than not, however, these creative acts have actually arisen "as if by design." This revelation requires a much deeper dive into the histories of these innovations in order to gain a firmer grasp on the very nature of behavioral creativity. What emerges is an intricate web of causation involving three main factors: context, consequence, and coincidence.

By concentrating on the *process* rather than the *product* of innovation, I hope to elevate *behavior* to its proper place—at the very center of creative human endeavor—for it is truly behavior that produces the innumerable innovations that have captivated thinkers' imagination. Those most splendid theories, goods, and gadgets would never have come into being without the behaviors of their inventors. This book thus stands apart from the majority of other discussions of creativity which stress "genius" and "foresight": two notions that do little more than to name what really needs to be explained.

We admittedly remain far from fully understanding how creative behaviors originate and evolve. But, I firmly believe that we'll gain ground far faster by abandoning fruitless fictions such as genius and insight, and by focusing instead on what can be observed and investigated. I hope that this book represents a step in the right direction.

Setting the Stage

> All history is retrospective. We're always looking at the past through the lens of later developments. How else could we see it? We are ourselves, as subjects, among those later developments. It's natural for us to take events that were to a significant extent the product of guesswork, accident, short-term opportunism, and good luck, and of demographic and technological changes whose consequences no one could have foreseen, and shape them into a heroic narrative about artistic breakthrough and social progress. But a legend is just one of the forms that history takes.

Louis Menand, 2015, *The New Yorker*

Genius. Inspiration. Insight. Foresight. Without deeper inquiry, most people blithely accept that these are the dominant forces that foster game-changing innovations by highly celebrated heroes. That's largely because it's all too easy to tell tall tales without doing the hard work of exploring the factors that actually participate in advancing the human condition.

But, the insightful or eureka moment is a myth: an altogether naïve and fanciful account of human progress. Innovations actually arise from an intricate web of historical, circumstantial, and adventitious influences, as American writer Louis Menand so perceptively proposed in the above quote from his engaging article about the history of rock and roll.

However, don't be intimidated by such complex determinism: a simple law of behavior is actively at work in the creative process. The *Law of Effect* inescapably and mechanically strengthens actions that have succeeded in the past over actions that have either failed or been maladaptive. The Law of Effect knows no bounds. It operates in all realms of human endeavor: sports, the arts, politics, science, medicine, and technology.

Originally formulated in 20th century psychological science, the Law of Effect debunks the myth of creative genius and closely parallels Darwin's *Law of Natural Selection*. Indeed, this second "selectionist" law is every bit as controversial as Darwin's evolutionary theory.

From its naturalistic perspective, the Law of Effect asserts that the strongest creative forces lie *outside* rather than *inside* of us. Most pointedly, we don't deliberately, foresightedly, and intelligently design our own behavior; rather, our behavior is shaped by our own past history and by the unique circumstances in which we find ourselves. That behavioral shaping process is often a haphazard, trial-and-error affair with no obvious end in sight; yet, it can generate innumerable innovations that significantly advance the human condition and deceive historians into incorrectly interpreting them as due to flashes of creative insight or foresightful design.

The parallel between the Law of Effect and the Law of Natural Selection is keenly drawn in the following quotation by British author Anthony Gottlieb in his incisive article about evolutionary psychology. In it, Gottlieb also addresses the fraught nature of such terms as design, planning, and purpose, especially in connection with our own thoughts and actions:

> The coup of natural selection was to explain how nature appears to be designed when in fact it is not, so that a leopard does not need a God to get his spots. Mostly, it doesn't matter when biologists speak figuratively of design in nature, or the 'purpose' for which something evolved. This is useful shorthand, as long as it's understood that no forward planning or blueprints are involved. But that caveat is often forgotten when we're talking about the 'design' of our minds or our behavior.
>
> Anthony Gottlieb, 2012, *The New Yorker*

Does this analysis seem too abstract? Too academic? Too scientific? Well, it isn't. Although my ideas are grounded in basic psychological science, I'll make my case in this book by guiding readers through a wide-ranging collection of 25 illuminating vignettes where the actual origins of noteworthy innovations have been revealed. Occasional vignettes may fall somewhat short of that ideal; but, they too will shine additional light on cases in which innovations may fail to break through to the mainstream.

Rather than reporting the results of arguably dry and technical laboratory studies, I'll focus on real life examples to piece together a narrative that I believe will more effectively communicate the essence of those laboratory findings to nonscientists. Here, I'll follow the sage advice of Nick Enfield, who persuasively advocates the critical part played by story-telling:

> Science can't exist without telling a story. The question is not whether we should use it, but how we should use it best. Only with a story can the facts be communicated, and only then can they become part of the received knowledge that drives the very possibility of scientific progress.
>
> Nick Enfield, 2018, *The Guardian*

I believe these vignettes offer scant evidence of insight and foresight in the creative process. Those creations generally do not arise *by design*; rather they arise *as if by design*. Perhaps we should even bid fond farewell to design: *"bye design!"*

Those vignettes further disclose that creativity lies at the heart of *all* adaptive action. Although we rarely appreciate it, each of us innovates every day as we adjust to life's many trials. Large and small, those innovations enable us, as individuals and as a species, to thrive.

Nevertheless, the world that we have fashioned has unleashed new and daunting challenges, such as overpopulation, pollution, hyper-partisanship, and climate change. Those "unintended consequences" of societal and technological progress now prompt the greatest question humanity has ever faced: Can we innovate our way out of the very tribulations that we ourselves have created?

If we are to survive, then the answer had best be *yes*. But, guarded optimism requires that we humbly accept that there will be no immediately obvious and direct routes to solving many of these formidable problems: only meandering paths involving the same kind of trial-and-error process that created them can be expected to succeed. To believe otherwise would be to fall prey to another myth: that of human invincibility.

Before proceeding further, let me put some meat on these bones by reprising the case of the devastating 2010 BP oil spill.

THE ULTIMATE GAME OF TRIAL AND ERROR

By 2010, several salient misadventures had already made British Petroleum (BP) a company many love to hate. Therefore, it proved unsurprising that BP's numerous failed efforts to contain the Gulf oil spill provoked so much derision and that its eventual success in plugging the oil leak prompted so little praise.

You may recall much of the colorful lingo for BP's various plugging techniques: "top hat," "junk shot," "saw and cap," "top kill," "static kill," and "bottom kill." You may also recall both commentators and BP employees alike describing the enormity of BP's task. After the deadly destruction of the Deepwater Horizon, Paul Ausick remarked that: "The oil industry has gotten better at preventing and stopping leaks, but a disaster of the kind that BP is battling now is unprecedented. Trying to stop the leaks is essentially a trial-and-error process that is based on experience gained at shallower depths plugging far smaller leaks" (5/11/2010, *24/7 Wall St*). Eric Rosenbaum asked: "Is BP just buying time again, or is it just learning by unavoidable trial and error in a situation where it continues to fail, but deserves an 'A' for effort?" (5/13/2010, *TheStreet*). Michael Haltman contended that: "BP is now engaged in the ultimate game of trial and error" (5/24/2010, *Homeland Security Examiner*). And, perhaps most unsatisfyingly of all, BP executive Kent Wells humbly confided that "We will just learn as we go" (5/16/2010, *MSNBC*).

LESSONS TO BE LEARNED THROUGH ADVERSITY

BP's problem was undeniably extreme. In our daily lives, we confront far less dire difficulties. Nevertheless, in any challenging situation, the solution is necessarily hard to find. What to do? For starters, you try things that have worked before in similar situations. Yet, because the situations often differ, those solutions may have to be tailored to deal with the new realities. But, even these modest modifications may not work. Then what? You try new things. Many may fail outright. Others may fail, but show promise; such promising leads can then be tinkered with until they do succeed.

This stumbling, error-filled process is actually how we all learn. As American author Phyllis Theroux has claimed: "Mistakes are the usual bridge between inexperience and wisdom." Trial-and-error learning is therefore not to be underestimated and ridiculed, but instead to be understood and appreciated. Quick and easy fixes are rare. Hard work, tolerance to frustration, and persistence pay off. So, too, does openness to fresh approaches when familiar ones fail. "Mistakes are the portals of discovery," held Irish novelist James Joyce. Such painful lessons have been well learned, especially by America's most illustrious inventor, Thomas Edison, who claimed that, "every wrong attempt discarded is another step forward." More recently, American engineer and inventor of the Segway, Dean Kamen has noted that, "most of the problems we seek to solve require years, sometimes decades, of trial and error." Social and political advancements too can be bumpy affairs. The Pulitzer Prize winning presidential biographer Jon Meacham has exhorted that, "after we fail, we must try again, and again, and again, for only in trial is progress possible." And, perhaps most famously and ungrammatically, Yoda in *Star Wars: The Last Jedi* perceptively proclaimed, "The greatest teacher, failure is."

For over a century, behavioral science has intensely and assiduously explored trial-and-error learning, although most people know little of this work (interested readers can consult Schwartz, Wasserman, & Robbins, 2002 to learn more). The psychologist Edward L. Thorndike began studying trial-and-error learning at the end of the nineteenth century (Lattal, 1998). He watched hungry cats learn to escape from "puzzle" boxes with tidbits of fish as their reward for doing so. The cats often flailed about before tripping the switch that secured their release and receipt of reward. The amount of time it took for the cats to escape traced "learning curves" whose shapes were often irregular, but generally downward, testifying to increasingly efficient performance. Thorndike proposed that a powerful selection process had transpired: starting with random acts, unsuccessful ones ceased, whereas successful ones continued, much as Charles Darwin's better-known Law of Natural Selection eliminates unfit organisms and retains fit ones.

Thorndike named this process of selection by consequences the Law of Effect. Countless empirical studies have since confirmed that this law applies with equal force to animals and humans (Rosenbaum & Janczyk, 2019; Skinner, 1953). It even applies to creativity; rewarding improbable behavior has been found to engender still more improbable, some might say, creative behavior in people, porpoises, and rats (Schwartz et al., 2002). The Law of Effect has also produced important practical benefits, including: behavioral therapy, computer-assisted instruction, treatments for drug addiction, and brain-computer interfacing which permits amputees to control artificial limbs through brain activity.

But, being mechanically controlled by the Law of Effect does have its pitfalls—especially when short-term outcomes overshadow long-term consequences. President Obama (6/3/2010, *Larry King Live*) made just this point regarding the Gulf oil spill: "I am furious at this entire situation because this is an example where somebody didn't think through the consequences of their actions. It's imperiling not just a handful of people. This is imperiling an entire way of life and an entire region for potentially years."

Indeed, it is all too common for people to underestimate the later repercussions of their actions; foresight is notoriously shortsighted. A wealth of scientific research has discovered that both people and animals exhibit poor self-control: they often choose a small, immediate reward over a larger, delayed reward (Mischel, 2014). Actually, this irrational choice accords with a key provision of the Law of Effect: namely, that rewards are discounted when they are delayed (Rung & Madden, 2018). This effect of delay on reward value underlies irrational choices and procrastination (Wasserman, 2019).

What can we learn from the Gulf oil spill? The Law of Effect holds for everyday human behavior, not just for animals in science laboratories. Consequences shape our behaviors, especially those consequences that promptly follow behavior. The Law of Effect generally promotes adaptive behavior, retaining successful actions and eliminating unsuccessful ones. Yet, this decidedly trial-and-error process is neither rational nor infallible. Whatever its limits and liabilities, the Law of Effect provides our best means of surviving in a harsh and uncertain world.

A more studious discussion of the Law of Effect and behavioral innovation is presented at the end of the book following the illustrative vignettes I've assembled. After all, I'm an experimental psychologist. My own views have been shaped by 50 years of studying the learned behaviors of humans and nonhuman animals. I strongly believe that the powerful investigative methods that psychological scientists currently employ are absolutely essential to revealing our true nature and its relationship to other living beings. That said, it will not be necessary for readers to consult this later presentation. My message should also be effectively conveyed by the vignettes that follow.

REFERENCES

Enfield, N. (2018, July 19). Our job as scientists is to find the truth. But we must also be storytellers. *The Guardian*.
Gottlieb, A. (2012, September 10). It ain't necessarily so. *The New Yorker*.
www.newyorker.com/magazine/2012/09/17/it-aint-necessarily-so
Lattal, K. A. (1998). A century of effect: Legacies of E. L. Thorndike's *Animal Intelligence* monograph. *Journal of the Experimental Analysis of Behavior, 70*, 325–336.
Menand, L. (2015, November 15). The Elvic Oracle. *The New Yorker*.
www.newyorker.com/magazine/2015/11/16/the-elvic-oracle

Mischel, W. (2014). *The marshmallow test: Mastering self-control.* New York, NY: Little, Brown.

Rosenbaum, D. A., and Janczyk, M. (2019). Who is or was E. R. F. W. Crossman, the champion of the Power Law of Learning and the developer of an influential model of aiming? *Psychonomic Bulletin & Review, 26,* 1449–1463.

Rung, J. M., and Madden, G. J. (2018). Experimental reductions of delay discounting and impulsive choice: A systematic review and meta-analysis. *Journal of Experimental Psychology: General, 147,* 1349–1381.

Schwartz, B., Wasserman, E. A., and Robbins, S. J. (2002). *Psychology of learning and behavior* (5th ed.). New York, NY: W. W. Norton & Company.

Skinner, B. F. (1953). *Science and human behavior.* New York, NY: Macmillan.

Wasserman, E. A. (2019). Precrastination: The fierce urgency of now. *Learning & Behavior, 47,* 7–28.

Previewing the Vignettes

Over the past fifteen years, I've been collecting vignettes that interestingly illuminate the nature of behavioral innovation. As a psychological scientist concerned with how both humans and animals adjust to the frequently changing demands of survival, it is perfectly clear that the most creative of those behaviors have posed the greatest challenge to existing theories of learning and behavior.

Many years ago, B. F. Skinner and his students conducted a series of studies – the so-called *Columban Simulation Project* (described and discussed by Epstein, 1981, 1985) – which suggested that animals such as pigeons might also produce behaviors that casual observers would deem to be creative, ingenious, or insightful if humans had been the species under investigation (Shettleworth, 2012 considered a more expansive list of species in her discussion of "insight" in animals). More recently, Scarf and Colombo (2020) have further expanded the realm of the pigeon's cognitive capacities by providing additional evidence that we humans might not be the only clever organisms under the sun. And, Colin and Belpaeme (2019) have, with some success, applied a deep learning algorithm to simulate pigeons' learning of "insightful" problem solving (although pigeons' actual behavior may not quite make the grade for such designation; Cook and Fowler, 2014).

Basic to all of this and other basic scientific research is the idea that the Law of Effect might play a prominent, but not necessarily the sole, role in producing creative behaviors. Fresh or innovative behaviors might indeed be strengthened by their consequences – once they have happened. But, what about their initial occurrence? And, how might those innovative behaviors later evolve as conditions change? Finally, would readers feel comfortable extrapolating the results from such highly controlled laboratory investigations of pigeons and other animals to the daily lives of people?

So, as much as I admired the pioneering work of Skinner and Epstein as well as the research of subsequent behavioral scientists, I suspected that an altogether different approach was needed to illuminate the origins of creative behaviors. Human stories exemplifying the process of behavioral invention might just do the trick.

So, from a wealth of vignettes in my files, I've chosen twenty-five that I believe may persuade readers that an objective approach to the problem of creative behavior is possible – one that avoids invoking the empty notions of "genius" and "foresight" to fill the gap between ignorance and understanding. In their place, more clearly defined and objectively rooted factors are proposed (see the *Three Cs* ahead).

As to the vignettes themselves, they cover a wide range of human endeavor. Most of the people you will meet through these stories are not household names. Nor will you necessarily associate some of the names you do recognize with the behavioral innovations they produced. These facts should in no way diminish the importance of their accomplishments. Literally millions of lives have been saved as a result of some of these innovations. Important advances in science, sports, the arts, and culture have also followed. Key to their inclusion in this collection, however, is that we know a good deal about the actual determinants of these achievements.

THE THREE CS: CONTEXT, CONSEQUENCE, AND COINCIDENCE

Behavioral innovation is often said to be overdetermined: even singular effects can be determined by multiple causes. Virtually all of the collected vignettes reveal the importance of Context, Consequence, Coincidence. These Three Cs may each participate in behavioral innovations, but to varying degrees in any individual instance: One size most assuredly doesn't fit all. Let's consider each of these factors in turn.

Context. Everything happens at a particular time and in a particular place. This overall setting is frequently referred to as the prevailing *zeitgeist* – the defining character of a particular period of history as shown by the ideas and beliefs of the period. But, that context is usually much more limited when we consider the circumstances in which individuals find themselves. Furthermore, individuals have their own personal experiences that they bring to the historical moment. Laboratory research tries to minimize such conspicuous idiosyncrasies, but history celebrates and respects them – as we will here.

Consequence. Novel behaviors must have consequences if they are to take hold. The odds are decidedly against the strengthening of novel behaviors. Take the analogy of the forward pass in football. Two out of the three possible things that can result are unfavorable: an incompletion or an interception. Only the completed pass is a favorable outcome. The same goes for novel behaviors. Nothing good or something bad is likely to follow. Only a good outcome will reinforce a novel behavior. It's no wonder then that, under stable conditions, people respond in highly regular ways. It's only when uncommon circumstances arise that routine responses falter and novel actions emerge.

Coincidence. Finally, chance may bring people and possibilities together. Good fortune is what comes from taking full advantage of those possibilities. Of course, you can't plan on luck to get you out of a jam. Nevertheless, opportunities can arise from adversities. And, when they do, exploiting them makes the proverbial lemonade out of the lemons.

MAKING THE MOST OF THE VIGNETTES

Each of the vignettes tells the story of an innovation in behavior. All are very human stories; many are profound, even moving tales of personal achievement and disappointment. They are also extremely diverse, dealing with very different problems, personalities, and time periods. Yet, the participation of context, consequence, and coincidence is unmistakable in most of them. I'll try not to browbeat you by unduly highlighting these factors as I tell the stories; but, please do keep them in mind as you proceed through the individual vignettes in any order you choose. See if you don't agree with me that there are strong explanatory strands that bind all of these vignettes together. And, continually ask yourself whether or not the notions of "genius" and "foresight" add anything to your understanding of the creative processes these stories reveal.

REFERENCES

Colin, T. R. and Belpaeme, T. (2019). Reinforcement Learning and Insight in the Artificial Pigeon. *CogSci: Annual Meeting of the Cognitive Science Society*, 1533–1539.

Cook, R. and Fowler, C. (2014). "Insight" in Pigeons: Absence of Means–End Processing in Displacement Tests. *Animal Cognition, 17,* 207–220.

Epstein, R. (1981). On Pigeons and People: A Preliminary Look at the *Columban Simulation Project. The Behavior Analyst, 4,* 44–53.

Epstein, R. (1985). Animal Cognition as the Praxis Views It. *Neuroscience & Biobehavioral Reviews, 9,* 623–630.

Scarf, D. and Colombo, M. (2020). Columban Simulation Project 2.0: Numerical Competence and Orthographic Processing in Pigeons and Primates. *Frontiers in Psychology, 10,* 3017.

Shettleworth, S. J. (2012). Do Animals Have Insight, and What Is Insight Anyway? *Canadian Journal of Experimental Psychology, 66,* 217–226.

SECTION 2 **The Vignettes**

PART I **Sports**

1 Dick Fosbury's High Jump Was No Flop!

> Consider ... the Fosbury Flop, an upside-down and backward leap over a high bar, an outright – an outrageous! – perversion of acceptable methods of jumping over obstacles. An absolute departure in form and technique. It was an insult to suggest, after all these aeons, that there had been a better way to get over a barrier all along. And if there were, it ought to have come from a coach, a professor of kinesiology, a biomechanic, not an Oregon teenager of middling jumping ability. In an act of spontaneity, or maybe rebellion, he created a style unto itself.
>
> Hoffer, 2009, *Sports Illustrated*

Perhaps the clearest and most compelling case of a behavioral innovation arising without planning and foresight is the now-famous *Fosbury Flop*. Along with millions of other television viewers, on October 20, 1968, I watched in utter amazement as Oregon State University's twenty-one-year-old Dick Fosbury revolutionized the sport of high jumping with a gold medal and Olympic record bound of 7 feet, 4 1/4 inches at the Mexico City games (for video, Olympic, n.d.). Fosbury accomplished this fabulous feat by sailing over the crossbar head first and backward (Figure 1.1)! As colorfully described that day by the *Los Angeles Times*'s legendary sportswriter Jim Murray, "Fosbury goes over the bar like a guy being pushed out of a 30-story window."

How did this most unorthodox maneuver originate? Origin stories are notoriously difficult to corroborate. The people responsible are usually deceased and written records of the events involved are scant or lacking. However, Dick Fosbury has been quite available and willing to tell his tale, as he has on many different occasions (some excellent stories and quotations are to be found in pieces authored by Burnton, 2012; Cummings, 1998; Hoffer, 2009; Trower, 2018; and Turnbull, 1998; as well as in a 2014 interview in Spikes).

FIGURE 1.1 Dick Fosbury using his signature "Fosbury Flop" in competition on the Oregon State University campus.
Courtesy of the Beaver Yearbook Photographs (P003), Special Collections and Archives Research Center, Oregon State University Libraries

At first, Fosbury joked with sportswriters, boasting to some that because of his advanced university background in physics and engineering he had initially designed the Flop on paper, and telling others that he had accidentally stumbled on this technique when he once tripped and fell backward on his takeoff (Hoffer, 2009).

However, in later interviews, Fosbury provided a more candid and accurate account of the Flop's development, revealing that it actually unfolded over many years and involved countless trials and errors,

It was not based on science or analysis or thought or design. None of those things. Just intuition. It was simply a natural technique that evolved. The interesting thing was that the technique developed in competition and was a reaction to my trying to get over the bar. I never thought about how to change it, and I'm sure my coach was going crazy because it kept evolving. I adapted an antiquated style and modernized it to something that was efficient. I didn't know anyone else in the world would be able to use it and I never imagined it would revolutionize the event. I wasn't trying to create anything, but it evolved that way.

Fosbury further expanded on the birth of the Flop:

When I first learned to high jump at the age of 10 or 11, I tried jumping with the "scissors" style [see Heinrich Ratjen (Anonymous, 1937), performing the scissors high jump]. I used that style until I went into high school in Medford, Oregon, when my coach, Dean Benson explained that I would never get anywhere with that technique. He started me with the "belly roll" or straddle technique [see Esther Brand (Anonymous, 1952), performing the straddle high jump]. However, I was really lousy with that style. I expressed my frustration to coach and he said that if I really wanted, I could still use the scissors. So, I decided on the bus trip to the next meet (the Rotary Invitation in Grants Pass) to go back to the scissors. During the competition, as the bar was raised each time, I began to lift my hips up and my shoulders went back in reaction to that. At the end of the competition, I had improved my best by 6 inches to (5'10") and even placed third! The next 2 years in high school (and later in college), with my curved approach, I began to lead with my shoulder and eventually was going over head first like today's Floppers.

In this way, the Flop evolved, not from premeditated design, but from an entirely trial-and-error process that combined repeated effort with the biomechanics of Fosbury's gangly 6'4" physique. *Sports*

Illustrated writer Richard Hoffer (2009) incisively described Fosbury's subsequent efforts to refine his jump in this way: "It was on-site engineering, his body and mind working together, making reflexive adjustments with only one goal, getting over the bar. His arms and legs were still all over the place, but what looked like an airborne seizure was actually Darwinian activity. Those tics and flailings that served to get him even a quarter inch higher survived. The rest were gradually pared away." Hoffer's Darwinian analysis of "survival of the fittest" *behavior* is especially to be noted.

As for his later training and Olympic triumph, Fosbury remarked, "I didn't train to make the Olympic team until 1968. I simply trained for the moment. I never even imagined I would be an Olympic athlete. It always seemed to evolve."

What did Fosbury think of the seeming awkwardness of his Flop? After all, when he first competed with the Flop, competitors and spectators alike teased and derided him. "I believe that the flop was a *natural* style and I was just the first to find it. I can say that because the Canadian jumper, Debbie Brill was a few years younger than I was and also developed the same technique, only a few years after me and without ever having seen me."

Indeed, Brill did enjoy a successful high jumping career using her own rather similar method – the *Brill Bend* (see Debbie Brill's high jump in a photograph by Kommer, 1972). She placed first in four significant international competitions: the 1970 and 1982 Commonwealth Games, the 1971 Pan American Games, and the IAAF World Cup in 1979. Brill also competed in two Olympic Games, placing eighth in 1972 and fifth in 1984 (Verschoth, 1971).

A striking coincidence? Yes, indeed. But, perhaps not as striking as the fact that Bruce Quande – a student at Flathead High School in Kalispell, MT – was photographed by the *Missoulian-Sentinel* on May 24, 1963, at Montana's state high school meet flopping backward over the crossbar! This was the same month that Fosbury recalls having flopped for the first time at the Rotary Invitation track meet in Grants Pass, Oregon.

Giving credit where credit is due, Quande must be deemed the first Flopper, as he had been using the backward technique some two years prior to this 1963 meet. Nevertheless, Quande never enjoyed much success during high school or in his later high jumping career at St. Olaf College. Said Fosbury after learning of Quande's earlier efforts, "I think it's real interesting. Our stories sound parallel. This will be an historical asterisk." And, so it is.

An important additional factor must be acknowledged in the evolution of the Fosbury Flop – the landing pit. As Hoffer (2009) pointedly observed, "completing the Flop successfully was only half the battle; the return to earth still had to be negotiated. Few would even consider such an experiment in flight knowing they'd have to land on their necks."

When Fosbury was jumping as a sophomore in high school, he had to land in pits that were filled with wood chips, sawdust, or sand. On one occasion, Fosbury hit his head on the wooden border of an old sawdust pit. On another occasion, he landed totally out of the pit, flat on his back, knocking the wind out of him. The next year, Fosbury's high school became the first in Oregon to install foam rubber in its high jump pit, thereby cushioning the jumper's fall and encouraging the use of the potentially dangerous Flop. The Fosbury Flop and cushioned landing areas thus appear to have fortuitously co-evolved.

Beyond the historic achievement of the technique itself, the alliterative Fosbury Flop moniker has been synonymous with the high jump since Fosbury struck gold in Mexico City over fifty years ago. The origin of that nickname is also of interest (Trower, 2018),

> I'm very proud that I received the naming rights. But the term by which the style is known did not appear overnight. To tell the truth, the first time that I was interviewed and asked "What do you call this?" I used my engineering analytical side and I referred to it as a "back lay-out." It was not interesting, and the journalist

didn't even write it down. I noted this. The next time that I was interviewed, that's when I said "well, at home in my town in Medford, Oregon, they call it the Fosbury Flop" – and everyone wrote it down. I was the first person to call it that, but it came from a caption on a photo [in my hometown newspaper, the *Medford Mail-Tribune*] that said, "Fosbury flops over bar." The context in Oregon was that our town was on a river, very popular for fishing, an hour from the Pacific Ocean. And when you land a fish on the bank, it's flopping. That's the action, and so it's a good description by a journalist, and I remembered and adapted it.

Dick Fosbury not only invented the Flop, but he named it too!

Given the common goal of getting over the high bar, Fosbury broke ranks with his competitors and adopted a most unorthodox style. All in all, and in the words of Frank Sinatra's signature song, Fosbury might properly say, "I did it my way!" Yet, he would positively *not* say, as does the song, "I planned each charted course, each careful step along the byway!" Nothing could have been more contrary to the origins of Fosbury's signature style.

REFERENCES

Anonymous (1937). Scissors High Jump by Heinrich Ratjen in Berlin in 1937 [Photograph]. Retrieved October 19, 2020, from https://commons.wikimedia.org/wiki/File:Bundesarchiv_Bild_183-C10378,_Hermann_Ratjen_alias_%22 Dora_Ratjen%22.jpg

Anonymous (1952). Straddle High Jump by Esther Brand at the 1952 Helsinki Olympics [Photograph]. Retrieved October 19, 2020, from https://it.wikipedia.org/wiki/Esther_Brand#/media/File:Esther_Brand_1952.jpg

Burnton, S. (2012, May 8). 50 Stunning Olympic Moments: No28: Dick Fosbury Introduces 'the Flop.' *The Guardian*. www.theguardian.com/sport/blog/2012/may/08/50-stunning-olympic-moments-dick-fosbury

Cummings, R. (1998, May 23). A Quantum Leap Backward. *The Missoulian*. https://missoulian.com/a-quantum-leap-backward/article_36a8f0f2-872b-5ce9-8467-a80be50ef07a.html

Hoffer, R. (2009, September 14). The Revolutionary. *Sports Illustrated*. https://vault.si.com/vault/1004226#&gid=ci0258bf761003278a&pid=1004226—062—image

Kommer, R. (Photographer). (1972). Debbie Brill Performing the Brill Bend in 1972 in Essen, Germany [Photograph]. Retrieved October 19, 2020, from https://commons.wikimedia.org/wiki/File:Debbie_Brill_1972_b.JPG

Olympic (n.d.) *Dick Fosbury Changes the High Jump Forever – Fosbury Flop- Mexico 1968 Olympics* [Video]. YouTube. Retrieved October 19, 2020, from www.youtube.com/watch?v=9SlVLyNixqU

Spikes (2014, November 28). Floppin' Heck! [Interview with Dick Fosbury]. https://spikes.worldathletics.org/post/dick-fosbury-tells-spikes-why-its-called-the

Trower, J. (2018, October 19). Fosbury: A Beautiful Mind? *Medford Mail Tribune*. https://mailtribune.com/sports/community-sports/fosbury-a-beautiful-mind

Turnbull, S. (1998, October 18). Jumper Who Turned Top of the Flops. *The Independent*. www.independent.co.uk/sport/athletics-jumper-who-turned-top-of-the-flops-1179100.html

Verschoth, A. (1971, February 22). She Gets Her Back Up. https://vault.si.com/vault/1971/02/22/she-gets-her-back-up

2 Shedding Fresh Light on the History of the Butterfly Stroke

> It was only a matter of time before arms met legs, but when exactly it occurred is difficult to say.
>
> Marie Doezema, 2016, *The New Yorker*

Michael Phelps is arguably the greatest Olympic athlete of all time. He is unarguably the most decorated. Over the course of his extraordinary five-game Olympic swimming career, Phelps earned a total of twenty-eight medals: twenty-three gold, three silver, and two bronze. The butterfly was his signature stroke. In this particularly demanding event, Phelps individually earned six gold and six silver medals in the 100- and 200-meter races.

Those unfamiliar with the butterfly stroke are sure to find it baffling. The windmill motion of both arms, first flung upward and forward out of the water, and then thrust downward and backward into the water, coupled with the rhythmically undulating torso and powerful leg kicks, create a seemingly chaotic concatenation of strenuous bodily maneuvers. Perhaps because the butterfly is considered to be the most aggressively athletic of all swimming strokes (and is thus sometimes called the "beast"), it's rumored to be Russian President Vladimir Putin's favorite, thereby flaunting his manliness.

No less baffling than the stroke's peculiarity is its origin. As with most origin stories, the history of the butterfly stroke is a rather frustrating evolutionary tale to tell (Buchanan, 2017). The patchy story line involves a host of disconnected contributors, anecdotes of dubious authenticity, and a wealth of unresolved controversies, prompting one recent commentator (Doezema, 2016) to lament its "murky" provenance. Further contributing to the story's complexity is the fact that there are two historically unrelated elements of today's butterfly stroke: the flying or over-the-water arm stroke and the fish-tail or dolphin leg kick. Finally, there is the fact that the full butterfly stroke did not suddenly emerge; it gradually grew out of the already familiar but far less flamboyant breaststroke (Barney and Barney, 2008).

THE STROKE

As far as the double over-the-water arm stroke is concerned, many writers credit its invention to Sydney Cavill. This Australian swimmer emigrated to the United States, where he coached several competitors at San Francisco's Olympic Club. Also adopting the over-the-water arm stroke was German swimmer Erich Rademacher, who competed in the United States in 1926 and 1927 as well as in the 1928 Olympic Games in the Netherlands. In some of these breaststroke events, Rademacher has been claimed to have incorporated a single over-the-water arm stroke as he approached the turns and at the end of the race. Occasional use of the over-the-water or "fly-away" technique is also credited to the Spence brothers, Wallace and Walter, both of whom trained and coached at the Brooklyn YMCA.

However, most often discussed in connection with the fly-away arm stroke was American swimmer Henry Myers of Brooklyn's St. George Dragon Swim Club. Myers used the fly-away stroke for the entire breaststroke leg of a three-stroke medley race at a 1933 YMCA competition in Brooklyn. His success in that race and in later events convinced Myers – and his competitors – that this innovation could greatly enhance swimmers' breaststroke speeds. Myers also perceptively suspected that this double arm motion might inspire much greater interest in this aquatic event, as even he found it rather unexciting to watch a breaststroke race. Adding the fly-away – with its spectacular splashing and violent arm motion – to the breaststroke might be far more likely to spark fan fervor.

THE KICK

Moving next to the dolphin kick, its invention is occasionally credited to Jack Stephens, who is claimed to have executed the maneuver around 1907 at a public swimming bath in Belfast, Northern Ireland. However, this account appears to have been a Wikipedia hoax (Bartlett, 2015).

Other authors have been intrigued by the possible role that American Volney C. Wilson played in developing the underwater dolphin kick. Wilson is said to have explored its possibilities before beginning work on nuclear fission and the atomic bomb in the Manhattan Project. Wilson was a strong swimmer and an alternate on the 1932 Olympic water polo team who was allegedly inspired to explore this technique by his informal studies of fish propulsion at Chicago's Shedd Aquarium.

PUTTING IT ALL TOGETHER

All of that colorful history notwithstanding, the full realization of the dolphin kick and its successful integration with the over-the-water arm stroke is truly an Iowa story. The first public notice of Iowa's contribution to this evolutionary tale came in the October 1936 issue of *Esquire* magazine. In *Frog, Butterfly and Dolphin*, G. Clifford Larcom, Jr. teased readers with the titillating byline: "Traditional strokes go the way of bloomer bathing suits as the engineers revise swimming." He then proceeded to describe several current developments in swimming techniques. Most noteworthy among these developments were the dramatic modifications then underway in the breaststroke. Here is Larcom's colorful narrative of the Iowa story,

> The latest breast stroke creation, a muscle devasting affair, is the dolphin stroke, modelled by Jack G. Sieg and tailored by Coach David A. Armbruster, both of the University of Iowa. Mr. Armbruster's coach's eye was caught one day by the sight of young Mr. Sieg rushing along under water with no other means of propulsion than the undulating, wriggling motion of a fish. Alert Mr. Armbruster incorporated this type kick with the double overarm and the results were sensational for the good old breast stroke.

> It was like taking a bloomer girl from the [1890s] and stripping her of her ten to fifteen pound bathing costume comprising among other things that barrier reef of chastity, the swimming corset, a mighty straight jacket of canvas, steel and rope, and encasing her instead in any up to date affair of silk.
>
> The stroke completely junks the old breast stroke kick and eliminates the checking forces developed in the old method ... If sanctioned, it would correct the one factor that retards interest in the ordinary breast stroke – its slowness.

Now, with the help of previously undisclosed materials from The University of Iowa Department of Intercollegiate Athletics and the studious research of David E. and Robert K. Barney (2008), I can further expand on Larcom's rather ornamented account. The history of the butterfly stroke should be murky no more!

Contributing to the development of the butterfly stroke were the advanced physical facilities that were available to the University of Iowa swimmers. Those facilities were detailed by Iowa City's official historian Irving B. Weber (1979) – Iowa's first all-American swimmer in the 150-yard backstroke in 1922. On October 25, 1925, the Iowa State Board of Education authorized construction of the world's largest fieldhouse, including the world's largest indoor pool (Figure 2.1). Construction of the facility took a mere twelve months: from December 1925 to December 1926. A 20-yard pool in the campus Armory was also available for conducting controlled swimming experiments. It provided facilities for underwater photography and motion picture recording. In and of themselves, those facilities might have given Iowa swimmers a modest edge in training. They might also have attracted some of the nation's best swimmers to the Iowa City campus. But, it was Iowa's famed coach that most surely contributed to swimming history.

David A. Armbruster (Figure 2.2) never participated in a competitive swimming race. Nevertheless, he was The University of Iowa's first swimming coach, serving for forty-two years (1916–1958)

FIGURE 2.1 Fieldhouse pool soon after opening in the late 1920s. Author: Frederick W. Kent. Courtesy of the F. W. Kent Collection, University Archives, The University of Iowa Libraries

and compiling over 100 All-American honors. He coached fourteen NCAA champions as well as both gold medal (Wally Ris) and silver medal (Bowen Stassferth) winners in the Olympic Games. Armbruster served as President of the College Swim Coaches Association of America (CSCAA) in 1938 and was inducted into the International Swimming Hall of Fame in 1966. Nevertheless, he is perhaps most famously credited with inventing the butterfly stroke – the most recent swimming stroke for national and international competition, joining the freestyle, breaststroke, and backstroke – as well as originating the flip turn. Armbruster was widely recognized as the scientific "dean" of competitive swimming from the 1930s through the 1950s. He earned that esteemed reputation because of his careful and detailed studies of swimmers' various techniques, stimulated in part by his colleague and occasional collaborator C. H. McCloy. America's first "giant" in the bioscience of physical education and exercise,

FIGURE 2.2 Coach David A. Armbruster in the 1940s.
Author: Frederick W. Kent. Source: University of Iowa Archives. Courtesy of the F. W. Kent Collection, University Archives, The University of Iowa Libraries

McCloy was a staunch advocate of strength training (which Armbruster adamantly rejected for his swimmers believing that it would make them "muscle bound").

Coach Armbruster had, in 1911, seen George Corsan, Sr., one of his early instructors, demonstrate the fish-tail kick at a swim carnival in Corsan's hometown of Toronto, Canada. Later, in 1916, Armbruster attended another of these swimming carnivals, in which various entertaining stunts were performed involving imitations of different animals. In it, Corsan performed a "butterfly" stroke by swimming the breaststroke kick and fluttering his hands at his sides on the *surface* of the water, not *above* the water. That stunt bore little resemblance to today's butterfly overarm stroke, yet its name may have subliminally registered with Armbruster.

Most importantly, at a subsequent swimming exhibition, Armbruster witnessed a young boy perform what he called the "Italian Crawl": the double overarm pull combined with the standard breaststroke frog kick. Although unsanctioned by the Amateur Athletic Union, that arm motion was to be used in many later exhibitions and, after certification, in many later competitions. The seeds may thus have been sown for future breaststroke development.

While coaching at Iowa, Armbruster had empirically determined that performing the breaststroke with the butterfly pull and the orthodox frog kick proved to be a poor mechanical combination. The kick was actually a retarding action compared to the faster, more powerful action of the flying arms. To take full advantage of the increasingly popular overarm stroke, something had to change. But, what?

This is how Armbruster described the way, beginning in 1932, he went about connecting and integrating the overarm stroke with the dolphin kick (reported in Armbruster and Sieg, 1935),

> One day in a moment of relaxation and play, I saw Jack Sieg go under water, lie on his side, with his arms trailing at the sides, imitating a fish, imitating the undulating movement with his head. I have often seen boys do this in water but never saw anyone derive the speed that Sieg was able to attain from it. We then tried it with the body face down, and the result was even greater. We then had him do it for speed against some of our best flutter-crawl kickers— no one could beat him. This was very impressive, to say the least. He then tried the double over-arm recovery of the breast stroke using this kick with it for several strokes. The leg rhythm was a natural movement and adapted itself perfectly to the rhythm of the double over-arm recovery. We then started to train for longer distances and adjust the breathing in order to cover one hundred yards. Several weeks practice brought results of greater speed, but at the cost of greater energy output. The stroke is an exhausting one.

The team of Armbruster and Sieg eventually called the innovative merger of these two techniques – the butterfly overarm recovery plus two dolphin kicks – the "dolphin butterfly breast stroke" (Figure 2.3).

PUTTING IT ALL TOGETHER 35

FIGURE 2.3 Birthplace of the Butterfly. Banner hanging above the Olympic pool on The University of Iowa campus.
Author: Ed Wasserman

But, their work was far from finished. Because the entire kick is performed underwater, Armbruster conducted additional mechanical analysis with the aid of slow-motion pictures, taken both from five windows below the surface of the water and from above the Armory's smaller 20-yard pool. Further experimentation produced several additional refinements, which in turn yielded still greater boosts in speed while better conserving the energy of the swimmer.

MAKING IT OFFICIAL

Unfortunately, the new butterfly stroke generated considerable controversy because it did not comply with prevailing breaststroke rules, leading to its painfully slow acceptance by the competitive swimming community. Beginning that saga, and by special permission, the full butterfly stroke was first used in the medley relay in a dual swim meet held in the fieldhouse pool with the University of Wisconsin on February 25, 1935. The Iowa team included Dick Westerfield swimming the backstroke leg, Sieg swimming the butterfly in place of the standard breaststroke, and Adolph Jacobsmeyer swimming the free style. Sieg's time was some 5 seconds faster than the best previous 100-yard breaststroke!

Soon thereafter Armbruster attempted an even more persuasive demonstration. Here is how Barney and Barney (2008) described that event:

> In March 1935, at the NCAA Championship meet at Harvard University, Armbruster and other members of the NCAA Rules Committee gathered on the deck of the Harvard pool for an exhibition of, what Armbruster termed then, the Dolphin Breaststroke. His demonstrator, of course, was Jack Sieg. The committee members were impressed with the demonstration but failed to agree on any alteration of the college rules to include yet another variation of swimming breaststroke. Undaunted, Armbruster continued to press the issue. The following month, April 1935, in a lengthy article in the *Journal of Health and Physical Education*, Armbruster enhanced his narrative with photographs featuring five silhouette drawings conceptualized from moving pictures of Sieg demonstrating different phases of the stroke.

Alas, those concerted efforts and many more over the ensuing years proved to be unsuccessful. Only in 1952 did the Fédération Internationale de Natation (FINA) sanction the butterfly stroke as a new event. Finally, in 1956, it was added to the Olympic Games in

Melbourne, Australia as a separate competition. Of course, the butterfly stroke has been considerably refined and developed in the ensuing years; it is now the second fastest stroke in the water after the freestyle. And, to most swimming fans, it is the most exciting.

While Armbruster and Sieg were developing the butterfly stroke, these two innovators were also investigating a new and faster turn – they called it the "tumble" or "flip turn." Bob Barney, Olympic historian, recalls that Armbruster and Sieg's experiments with the tumble turn found it to be "unmechanical" and "unsuitable" for the breaststroke and butterfly breaststroke; but, the tumble turn did effectively integrate with the freestyle stroke (personal communication; August 28, 2020). In 1938, Armbruster and Sieg believed they had perfected the maneuver and used it for the first time in the NCAA Championships at Rutgers University. This was and still is the fastest turn in the water. It is currently used by all speed and distance swimmers.

Writing a personal letter to Buck Dawson (Executive Director of the Swimming Hall of Fame) on September 19, 1968, after being named the 1966 Honor Coach, David Armbruster confessed that, "As I ponder over all of [my] awards and thrills, I will choose the creation of the Dolphin Butterfly stroke and the flip turn as giving the greatest source of satisfaction. These two creations will live long, and beyond my time in the swimming world." So they have!

Returning to Larcom's entertaining *Esquire* article, one more point should be stressed. Namely, how these important evolutionary changes in swimming actually came about: "These latest in streamlined swimming strokes are developing out of their incipient stages. Speed becomes greater because waste motions have been discovered and eliminated and the strokes have been polished to smooth precision. [Although other strokes have continued to evolve] it is the breast stroke ... that has had the most universal development."

As in countless other areas of human endeavor, trial-and-error assumes center stage in the unfolding of behavioral innovation. No one – not even Armbruster or Sieg – could have envisioned the final result of their extensive aquatic experiments. Indeed, the success of their combining butterfly arms with dolphin kicks had to wait over twenty years to be officially sanctioned. Today's butterfly swimmers – including Michael Phelps – may have little knowledge of the stroke's origins. But, if we are to gain a proper appreciation of this significant sporting achievement, its history is essential. Given that extended history, it is obvious that today's butterfly stroke is no "stroke of good luck!"

REFERENCES

Armbruster, D. A. and Sieg, J. G. (1935). The Dolphin Breast Stroke. *The Journal of Health and Physical Education*, 6:4, 23–58. https://doi.org/10.1080/23267240.1935.10620880

Barney, D. E. and Barney, R. K. (2008). A Long Night's Journey into Day. *Journal of Olympic History*, 16, 12–25.

Bartlett, J. (2015, April 16). How Much Should We Trust Wikipedia? *The Daily Telegraph*. www.telegraph.co.uk/technology/wikipedia/11539958/How-much-can-we-trust-Wikipedia.html

Buchanan, J. (2017, May 25). The Butterfly: A Complex History for a Complex Stroke. *Swimming World*. www.swimmingworldmagazine.com/news/the-butterfly-a-complex-history-for-a-complex-stroke/

Doezema, M. (2016, August 11). The Murky History of the Butterfly Stroke. *The New Yorker*. www.newyorker.com/news/sporting-scene/the-murky-history-of-the-butterfly-stroke

Larcom, G. C., Jr. (1936, October). Frog, Butterfly, and Dolphin. *Esquire*. www.ishof.org/assets/1936-history-of-swimming-stokes.pdf

Weber, I. B. (1979, September 22). Too Bad the Fieldhouse Pool Can't Tell Its Story. *Iowa City Press Citizen*.

3 The Monkey Crouch
Jockeying for Position

> It may be insular, narrow-minded, prejudiced, and the rest of it, but I cannot believe that, generation after generation, jockeys have been sitting on the wrong part of a horse's back, that the best place for the saddle is not where it has always been, and that at the end of the nineteenth century the theory and practice of horsemanship as applied to racing is to be revolutionised.
>
> Alfred E. T. Watson, 1899, *Badminton Magazine*

We often take things for granted. Things as they are now must always have been this way, right? But, things really do change – sometimes in dramatic ways. Such is the case with the position of the jockey on a thoroughbred racehorse. The crouched style to which we are so firmly accustomed was actually a drastic departure from the upright riding style (see an illustration by Cameron, 1872) that survived until the very beginning of the twentieth century.

Recent interest in this striking modernization of race riding was sparked by a highly publicized report that documented the superiority of the stooped posture over the erect posture (Pfau et al., 2009). The so-called monkey crouch confers considerable locomotor benefits for the horse as well as affording a smaller reduction in wind resistance for the horse and rider (Holden, 2009).

This novel riding style was likely to have produced the marked boosts in speed that were recorded by the race winners at the English Epsom Derby Stakes from 1900 to 1910 – an interpretation supported by careful physical measurements (Pfau et al., 2009). Of course, a century earlier, these methods of measurement were unavailable to thoroughbred jockeys or trainers. So, how might this new riding style have originated? Herein lies the interesting and interrelated tale of two tales told by two contemporaneous horsemen.

According to the first, generally accepted tale, the originator of the monkey crouch was Tod Sloan (1874–1933; see his riding style caricatured by Giles, 1899). Sloan was a truly flamboyant character, whose eventful life was carefully chronicled by John Dizikes in his biography, *Yankee Doodle Dandy: The Life and Times of Tod Sloan* (2000).

James Forman Sloan was born in Bunker Hill, Indiana. James was rather callously dubbed "Toad" as a child, an epithet he gladly changed to "Tod." Sloan had a hardscrabble upbringing. His mother passed away when he was only five years old and his father abandoned him to the care of neighbors. By the time he was thirteen years old, Sloan was pretty much on his own, doing menial jobs, laboring in oil fields, and rambling with sideshows from St. Louis to Kansas City. His diminutive size seriously restricted his occupational opportunities. Even in his twenties, Sloan was under five feet tall and weighed only 90 pounds. His small size went from being a liability to an asset when, after working as a stable hand, he followed his brother's lead and began riding race horses.

Sloan was initially a lackluster jockey, experiencing considerable fear when mounted on horseback. Nevertheless, he persisted in this pursuit and gained valuable proficiency racing first in Chicago in 1888 and later in northern California in 1892 and 1893. His numerous victories in San Francisco inspired him to compete at more prestigious New York tracks. By 1896, Sloan's winning ways thrust him into the limelight and made him a national celebrity, albeit one of some ambivalence owing to his overdeveloped sense of self-importance.

Sloan journeyed to England in 1897 and 1898 to expand the scope of his American reputation. Expand it he did! On October 2, 1898, *The New York Times* led with the headline: "Tod Sloane's Great Success on the English Turf Has Astonished the Old World. FIVE WINNERS IN ONE DAY." The accompanying story elaborated,

> 'Tod' Sloane, the American jockey, has during the past few years been a very successful jockey. Before he became popular he was considered a very ordinary rider, but now Princes, Dukes, Earls, and members of Parliament are clamoring for his services. He has astonished the English turfmen and jockeys not only by the style of riding which is peculiar to him, but also by his great success. . . . Sloane sits in the saddle crouched well forward on his mount's neck, and in the position there may be found a partial explanation of his success. His body in this position offers less resistance to the wind than that of a jockey who rides bolt upright. (20)

Sloan's staggering success was trumpeted by the British press as well. On October 12, 1898, a full-page story appeared in *The Sketch*, a newspaper published by the *Illustrated London News*. It unreservedly proclaimed him to be "The Greatest Jockey of the Day. Tod Sloan, the American, has come, seen, and conquered."

Beyond his racing success, the British were also smitten with Sloan's "Yankee brashness." However, by 1900, their infatuation had soured; his insufferable cheekiness and his heavy wagering provoked a stern injunction from the British Jockey Club prohibiting Sloan from racing because of "conduct prejudicial to the best interests of the sport." This injunction was also honored in America, thereby prematurely ending his riding career at the age of 26. Sloan's remaining years were remarkable for their abject aimlessness. He was twice married, became a vaudevillian, a bookmaker, a bar owner, and a bit actor in motion pictures. Sloan died at the age of 59 of cirrhosis of the liver in Los Angeles.

Perhaps the most memorable detail about Sloan is that he was the inspiration for George M. Cohan's famous 1904 musical, *Little Johnny Jones*, which featured the immensely popular song: "Yankee Doodle Boy." It was Tod Sloan who came to London, just to ride the ponies; he was the Yankee Doodle boy!

Of course, for our purposes, we are most concerned with how Sloan is believed to have developed the monkey crouch. It just so happens that, in 1915, Sloan penned his autobiography – *Tod Sloan: By Himself* – in which he recounts that tale,

> One day, when I and Hughie Penny, who was then a successful jockey, were galloping our horses to the post, my horse started to bolt, and in trying to pull him up I got up out of the saddle and on to his neck. Penny started laughing at the figure I cut, and I laughed louder than he, but I couldn't help noticing that, when I was doing that neck crouch, the horse's stride seemed to be freer, and that it was easier for me too. Before that I had seen a jockey, named Harry Griffin, riding with short stirrups and leaning over on the horse. As he was the best jockey of the day I put two and two together and thought there must be something in it, and I began to think it out, trying all sorts of experiments on horses at home. The "crouch seat," the "monkey mount," or the thousand and one other ways it has been described, was the result. Then the time came when I determined to put it into practice. But I couldn't screw up enough courage the first time I had a chance. I kept putting it off. At last, though, I did really spring it on them. Everybody laughed. They thought I had turned comedian. But I was too cocksure to be discouraged. I was certain that I was on the right track. I persevered, and at last *I began to win races!* (22–23)

From his telling, Sloan successfully parlayed a clumsily comical racing mishap into a new riding style by initially recalling having seen another jockey similarly positioned on his horse and by later conducting a series of practical trials before finally putting it into action at the racetrack. That's surely of great interest. But, Sloan's account omits any mention of another American jockey, Willie Simms (1870–1927), who may have invented the monkey crouch even before Sloan (Riess, 2011), although there seems to be no account describing how Simms did so.

Simms was a very successful jockey who was inducted into the U.S. Racing Hall of Fame after winning five of the races that would later constitute the Triple Crown. Simms too had traveled to England two years earlier than Sloan, where he experienced both initial success and mocking laughter. Simms won several races while adopting the "forward seat" style of racing; however, he was of African American ancestry and was not warmly embraced by the English racing fraternity, making his stay a brief one. Indeed, the very moniker "monkey crouch" may have originated from racial prejudice, as this entry from the London weekly newspaper *The Graphic* clearly reveals,

> Of greater interest than the horses [brought from the United States to race in England] has been the American rider, Simms. He rides very forward on the saddle, with shortened stirrup leathers that force his knees high up, and as he leans so much forward that his hands are within a few inches of the bit, he presents a living image of the monkey on a stick, the resemblance being heightened by his negroid cast of countenance. (May 11, 1895, 17)

The continued use of the term "monkey crouch" is indeed lamentable as it has perpetuated this vile slur. It is doubly unfortunate because, as Sloan noted in the prior quotation, a variety of other nicknames had already been given to the crouched racing posture. Among them, the "American seat" might be one to have been judiciously avoided, given the second tale of the style's origin. The teller of this tale could not have been more different from Tod Sloan.

Harding Edward de Fonblanque Cox (1854–1944; see his caricature published in Vanity Fair in 1909) was born in Bloomsbury, a fashionable residential district in London's West End and home to many prestigious cultural, intellectual, and educational institutions. Cox's father was a man of substantial wealth and high station: a lawyer, legal writer, publisher, landowner, and justice of the peace.

As a lad, Harding lived a life of great privilege. His secondary education was at Harrow School, an independent boarding school for boys, whose graduates include renowned poet Lord Byron, Prime Minister of Britain Winston Churchill, Prime Minister of India Jawaharlal Nehru, and actor Benedict Cumberbatch. From 1873 to

1877, Cox attended Cambridge University, where he earned his Bachelor of Arts degree and completed his legal training as a barrister.

Cox's father died in 1879 bequeathing him a great sum of money. Despite his inherited wealth, Harding did practice law; however, he also pursued diverse literary, dramatic, and journalist interests. Cox was especially engrossed in the theatre and it was here that he met his future wife, Hebe Gertrude Barlow, a singer and well-known actress. Cox was also an avid sportsman and excelled at shooting, cricket, rowing, and hunting.

Cox's passion for riding went hand in hand with his passion for hunting. He owned several racehorses and was an amateur jockey. Cox even obtained a special license from the Jockey Club and the National Horseplayers Championship to ride on equal terms with professional jockeys.

In 1922, Cox's journalistic and sporting interests came together in an engaging and freewheeling memoir, *Chasing and Racing*. In it, he endeavors to set the record straight as to the origin and nature of the monkey crouch. Cox writes,

> There is an erroneous impression extant that Tod Sloan was the first American jockey to exploit the "monkey crouch" in this country. As a matter of fact, it was introduced by a compatriot of his, one Simms, a mulatto or quadroon, who won several races and attracted considerable attention, owing to what was at that time considered an extraordinary seat. But even before Simms there was an inconspicuous amateur who adopted the "crouch" in a modified form. As late as 1921 a leading sporting paper published a par in words to the following effect:
>
> An old race-goer says "that Harding Cox had the worst seat and the best hands of any jockey he ever saw. Cox was crouching before Sloan, or even Simms, was ever heard of." And this is a fact. (211–212)

A fact? Really? Are we to believe that an utter unknown was racing with the monkey crouch before either Simms or Sloan arrived in England?

Cox has still more to say on his own behalf. He specifically details how he happened on the monkey crouch and what benefits the novel position conferred to race riding:

> When hunting, I rode very short, and leant well forward in my seat. When racing, I found that by so doing I avoided, to a certain extent, *wind pressure*, which ... is very obvious to the rider. By accentuating this position, I discovered that my mount had the advantage of *freer hind leverage.* Perhaps that is why I managed to win on animals that had been looked upon as "impossibles," "back numbers," rogues and jades. (212)

But, how much credence should we give to these observations from a gentleman hunter and amateur jockey? Let's see what more Cox has to say,

> My theory was endorsed by Tod Sloan himself. I had been brought into touch with the little man soon after his arrival in England, and, being greatly interested in his methods, I cultivated his acquaintance. (212)

Oh my, that's quite a coincidence! These two men of radically different upbringing but common interest actually met. Cox then proceeds to describe an extended conversation he had with Sloan, whom he colorfully called the "Wizard of the West",

> One day, when we were seated on the Terrace at Monte Carlo, we discussed the merits of the "crouch." Tod not only discoursed sweetly on the topic, but drew a spirited sketch of a race horse and its anatomy, explaining how the distribution of weight under his system helped the general action of the animal. [Sloan] went on:
> "Say, I figure that this seat has considerable advantage; but it's no cinch for any guy or dud jockey that takes a hand. Believe me, sir, there's nothing to it unless you have the whole bunch of tricks up your sleeve." (212–213)

Tricks, yes, but prohibitions, too, as further conversation explained,

> Sloan managed to win on horses which had erstwhile been regarded as only fit for cats' meat. Apart from his seat, he had the best of hands, and always seemed to be on most excellent terms with his mounts. The flail – as a means of exacting an expiring effort from a horse which had already exerted its last normal effort to win – was "off the map" as far as Tod was concerned, for he was a genuine lover of horse-flesh. (213)

There we have it. Two innovative horsemen, two intriguing tales. Sloan and Cox each incidentally stumbled onto the crouched position; they each pursued their serendipitous observations with trial-and-error investigations to improve their techniques; they each believed that there were distinct motor benefits to the freedom of the mount's stride; and, they each gained an advantage on their competitors by adopting this new style. Must we now credit two creators of the crouch?

I, for one, wouldn't be at all averse to doing so. However, others might very well dismiss both Sloan and Cox as having originated the monkey crouch.

Dizikes (2000), for example, takes a skeptical view of any single creator. "In a culture prizing individualism, the search was always to identify the one autonomous inventor" (67). Dizikes takes special aim at Sloan, hinting that in telling his tale, he may have embellished his involvement in the discovery and wondering whether Simms might have beaten Sloan to the punch. Furthermore, Dizikes raises doubts concerning Simms as the sole originator, instead suggesting that this "revolutionary development" may have resulted from an "evolutionary collective process" involving many individuals, including white and African American stable boys who exercised horses without suitable riding gear and Native Americans riding bareback high up the necks of their horses. Such an evolutionary process would have unfolded over an extended period of time, making it difficult if not impossible to pinpoint a time, place, or person as being critical to its origin.

In sports or in other competitive realms, innovations are unlikely to stay new for long. If they prove effective, then novelties quickly become established because rivals will adopt them in their quest to remain competitive. Nothing succeeds – or is so prone to be copied – like success!

That the monkey crouch has survived for a century as the predominant racing style is testimony to its effectiveness. To whomever credit belongs – Tod Sloan, Willie Simms, Harding Cox, or innumerable unknowns – it should be clear that the provenance of the monkey crouch was not due to foresighted design. The evolutionary processes of variation, selection, and retention seem to have been hard at work in its development as in so many other instances of behavioral innovation.

REFERENCES

Cameron, J. (Artist). (1872). *Standard Upright Racing Style Prior to the Introduction of the Monkey Crouch. Harry Bassett and Longfellow at Saratoga, New York, July 16, 1872 (and) at Long Branch, New Jersey, July 2, 1872* [Poster]. Retrieved October 19, 2020, from https://commons.wikimedia.org/wiki/File:Harry_Bassett_and_Longfellow_at_Saratoga,_N.Y.,_July_16th_1872_(and)_at_Long_Branch,_N.J.,_July_2nd_1872._LCCN2002695833.jpg

Cox, H. (1922). *Chasing and Racing: Some Sporting Reminiscences.* New York: Dutton.

Dizikes, J. (2000). *Yankee Doodle Dandy: The Life and Times of Tod Sloan.* New Haven, CT: Yale University Press.

Giles, G. D. (Artist). (1899). *Caricature of American Jockey Tod Sloan* [Illustration]. Retrieved October 19, 2020, from https://commons.wikimedia.org/wiki/File:Tod_Sloan_caricature.jpg

Holden, C. (2009, July 16). How the "Monkey Crouch" Transformed Horseracing. *Science.* www.sciencemag.org/news/2009/07/how-monkey-crouch-transformed-horseracing

Pfau, T., Spence, A. J., Starke, S., Ferrari M., and Wilson, A. M. (2009). Modern Riding Style Improves Horse Racing Times. *Science, 325,* 289.

Riess, S. A. (2011). The American Jockey, 1865–1910. *Transatlantica*, 2.

Sloan, T. (1915). *Tod Sloan: By Himself*. New York: Brentano's.

Vanity Fair (1909). *Caricature of Mr. Harding Edward de Fonblanque Cox*. Retrieved October 19, 2020, from https://commons.wikimedia.org/wiki/File:Harding_Edward_de_Fonblanque_Cox,_Vanity_Fair,_1909-09-01.jpg

Watson, A. E. T. (1899). Racing, Past and Future. *Badminton Magazine*, 8, 26–27.

FURTHER MATERIAL

Asleson, R. (n.d.). Tod Sloan: Jockeying to Fame in the 1890s. Face to Face: A Blog from the National Portrait Gallery. https://npg.si.edu/blog/tod-sloan-jockeying-fame-1890s

4   Riding Acey-Deucy

Win number 2,467 had been notched just two races earlier in the day. Now, win number 2,468 was a trifling 70 yards away. Veteran jockey Jack Westrope (variously nicknamed "Jackie," "Strope," and "the Rope") swung his race-favorite horse, *Well Away*, around the outside of hard charging *Midnight Date* and *Nushie*, and was galloping down the homestretch to what looked like certain victory in the seventh, final "feature" race of the day – the Hollywood Oaks – at Hollywood Park Racetrack, when his 3-year old filly unexpectedly veered left toward the infield rail (Hovdey, 2002). Westrope frantically took to the whip with his left hand in a last-ditch effort to redirect the errant steed, but to no avail; *Well Away* stumbled, violently flinging Westrope hard into the inside rail on his back. The time was 5:17 PM on June 19, 1958.

Displaying no external injuries, but limp and unconscious, Westrope was sped by ambulance to nearby Centinela Hospital where, although briefly regaining consciousness, the 40-year old American Hall of Fame rider died from internal injuries and uncontrolled bleeding 2 hours later. In the grandstands were Westrope's parents (William and Lotus, both Iowans I must note) for whom this tragedy wrenchingly replayed the death of their elder son. On February 27, 1932, they had witnessed Tommy perishing from a broken neck in another freak mishap at Tijuana, Mexico's Agua Caliente Racetrack (Simon, 2002; Smith, 1958).

Hollywood Park Racetrack was located only 1 mile from my boyhood home in Inglewood, California (the property now houses SoFi Stadium). Despite my close proximity to the track, I had no interest in thoroughbred racing and I was entirely unaware of Westrope's tragic death. I was then twelve years old and an avid fan of the newly arrived Los Angeles Dodgers major league baseball team, which had recently relocated from Brooklyn, New York, and had played its first game just two months earlier on April 18.

My unanticipated interest in Westrope was piqued more than five decades later on August 14, 2009, when I emailed my cousin Ron regarding my growing curiosity about the origins of innovative behaviors. Ron was a dedicated horseracing fan. Because of his considerable knowledge of the sport, he was able to alert me to an intriguing horseracing innovation, *acey-deucy*, which he suspected might be of relevance to my interest. Right he was! Just what is acey-deucy and what did Westrope have to do with its origin?

To best appreciate acey-deucy requires that one view the horse and jockey directly from the front or from the rear: the jockey's left stirrup iron is commonly placed from 2 to 12 inches lower than the right by individually adjusting the attached leather straps (also see Figure 4.1). Many horseracing experts believe that this so-called acey-deucy style of uneven stirrup placement confers important advantages on oval tracks, where only left turns are encountered in counterclockwise American races (ordinarily, English races are run clockwise, thereby reversing the relative lengths of the stirrup leathers to contend with right turns only). Although science has not yet confirmed the effectiveness of the acey-deucy racing style, knowledgeable observers believe that this practice permits the horse and rider to better lean into the turns, thereby providing the racing duo with better strength and balance, arguably the optimal formula for harnessing the centripetal force of a tight bend (Harzmann, 2002).

Many sportswriters have credited the two-time Triple Crown winning jockey Eddie (the "Master") Arcaro with perfecting and popularizing acey-deucy in the 1940s. In the 1957 *Sports Illustrated* series, *The Art of Race Riding*, Arcaro explained, "I find the short right iron gives me the great pushing action. I have the feeling I can get right down and shove on the horse with my right foot behind me and the left foot forward to shove against. Thus, my right foot is actually my balancing pole."

Jockey Myles Neff expanded on the merits of acey-deucy in his 2015 book, *Stylin': Reviving the Lost Art of Race Riding*,

FIGURE 4.1 Riding with the acey-deucy positioning of leathers and stirrups. The left leg is extended, the right leg is bent, and the jockey is "folded" over the horse.
Author: Myles Neff. In Stylin': Reviving the Lost Art of Race Riding. Shoot from the Hip (2015). Courtesy of Jason Neff

Establishing and maintaining the balance created by riding acey-deucy is the easiest and most efficient way for a racehorse and jockey to move together. It allows the rider to shift his balance and position to go with that of the horse and the horse's change of balance, whether it is small or large, gradual or quick. Acey-deucy allows the rider to "fold into" the horse instead of squat over him. (106)

However, Arcaro (1957) himself never took credit for this riding innovation; indeed, he professed to having no idea who coined the term. Actually, the most likely developer of riding acey-deucy was Arcaro's contemporary – none other than Jackie Westrope. Although a few other jockeys had been riding with the left stirrup iron a *little* lower than the right, Westrope dropped it *dramatically* lower. Some observers in fact reported that "[Westrope's] left leg was completely straight and his right knee was under his chin" (Harzmann, 2002).

Of course, questions about Westrope's innovation must be asked: Had the idea of riding acey-deucy occurred to Westrope in a sudden flash of insight? Had he fastidiously examined extensive film records with the ambition of devising a calculated plan to outrun his racing rivals? Had he conducted elaborate equestrian experiments to empirically assess the effectiveness of riding acey-deucy? Certainly not, we learn from one of Westrope's famous riding brethren.

Winner of eleven Triple Crown races Bill (nicknamed Willie and the "Shoe") Shoemaker – and the very jockey who guided the filly *Midnight Date* past the stumbling *Well Away* and the hurtling Westrope to win that tragic Hollywood Oaks – authoritatively reported just what prompted Westrope to take acey-deucy to such extreme lengths. "He did it after he came back from getting hurt one time, and he couldn't bend his left knee all the way. His balance was still good, though, and other guys started copying him" (Hovdey, 2002).

So, a gimpy left leg that resulted from a starting gate misadventure actually brought about Westrope's off-kilter racing style, which just happened to enhance a horse and rider's navigating left-hand turns. In other words, riding acey-deucy was a *doubly accidental* innovation! A racing *accident* damaged Westrope's left leg, which required him to place his stirrups at radically different heights. This uneven stirrup placement then *accidentally* supported faster racing performance because racing in the United States involves only left turns. How fortuitous it was that Westrope hadn't been racing in England!

Interestingly, a common saying may shed fresh light on the origins of this horseracing innovation: "Necessity is the mother of invention." One can plausibly suggest that riding acey-deucy might eventually have arisen as a natural consequence of the biomechanics of navigating tight turns. The jockey's need for speed and control might, by one means or another, have led to this inventive solution; after all, some riders are alleged to have been toying with this technique before Westrope's injury. Of course, the actual provenance of riding acey-deucy turns out to have followed a very different path. Nevertheless, in a most unorthodox way, necessity may again be said to have been the mother of invention, with Westrope's need to keep his left leg extended thus promoting the emergence of this still popular racing style (Hovdey, 2002).

One final note. Happy accidents like the origin of riding acey-deucy are often dubbed *serendipitous*. What was certainly not serendipitous was the swift and widespread adoption of this riding style by Westrope's racing competitors. A close-knit subculture such as that of jockeys is tailor-made for the rapid radiation of any new-found advantage, whether in sports or in business or in warfare. Taking all of these considerations into account, riding acey-deucy most assuredly cannot be attributed to foresight, planning, or design.

REFERENCES

Arcaro, E. (1957, June 24). The Art of Race Riding: Part 2 "Pre-Race," *Sports Illustrated*. www.racehorseherbal.com/Arcaro1.html

Harzmann, C. (2002, August 3). Jack Westrope: Quiet Little Man. *The Blood-Horse*. www.bloodhorse.com/horse-racing/articles/10803/Jackie-westrope-quiet-little-man

Hovdey, J. (2002, March 13). Hall Has Unfinished Business. *Daily Racing Form*. Retrieved December 25, 2012.

Neff, M. (2015). *Stylin': Reviving the Lost Art of Race Riding*. Osprey, FL: Shoot From the Hip Publishing.

Simon, M. (2002, August 3). Stars for All Time: Jackie Westrope. *Thoroughbred Times*.

Smith, R. (1958, June 17). Jackie Westrope didn't need help. *New York Herald Tribune*. http://news.google.com/newspapers?nid=2194&dat=19580627&id=0BAwAAAAIBAJ&sjid=V98FAAAAIBAJ&pg=6922,2411622

FURTHER MATERIAL

AP report, http://news.google.com/newspapers?nid=888&dat=19580621&id=ININAAAAIBAJ&sjid=UXYDAAAAIBAJ&pg=3963,191830

Historycomestolife (2011). *Tragic death of jockey Jack Westrope – 1958* [Video]. www.youtube.com/watch?feature=player_embedded&v=EEuh5Powpo0

Old timey race riding, http://racehorseherbal.net/racehorseherbal.com/blog/race-riding/old-timey-race-riding/

5 The High Five
One Singular Sensation

> Glenn Burke changed the world forever with that one hand slap.
>
> Jamie Lee Curtis

ONE FINE DAY

It was Sunday afternoon, October 2 and the sixth inning of the final game of the 1977 major league baseball season. Promising sophomore center fielder Glenn Burke stepped into the batter's box at the Chavez Ravine ballpark with his Los Angeles Dodgers and the visiting Houston Astros tied 2-2. Burke got the measure of opposing pitcher James Rodney Richard's first throw to the plate and spanked a home run, giving the Dodgers a 3-2 lead. It was also the *first* home run of Burke's major league career and it came following a protracted dry spell: 215 "at bats" in 108 games over 2 seasons (Newhan, 1977)!

CELEBRATE GOOD TIMES ... COME ON

This was certainly cause for celebration. After rounding the bases and stepping on home plate to validate his home run, Burke ran toward the Dodgers' dugout. As he was doing so, teammate Dusty Baker raced out to greet him and they gave each other the *second* recorded High Five in history (Mooallem, 2011). The *first* recorded High Five had actually happened mere minutes earlier.

With the count one ball and two strikes, left fielder Baker had just whacked a Richard fastball over the left centerfield wall at the 395-foot mark. That home run was Baker's thirtieth of the season and it set a major league record. Four Dodgers – Dusty Baker, Ron Cey, Steve Garvey, and Reggie Smith – had now each hit thirty or more home runs in a single season, a feat never before achieved in baseball history. This was the moment that the 46,501 fans in attendance had so eagerly anticipated, and they jubilantly cheered this extraordinary accomplishment by the Dodgers' quartet of sluggers, affectionately dubbed the "Big Blue Wrecking Crew." But, the fans could never have anticipated that they would also witness another event of even greater historical significance – the birth of the High Five.

Because this particular game was not televised, there is no video record of the *first* High Five. Nor is there even a photograph of that remarkable happening. However, a single blurry photograph is posted on the Internet that some believe captured that critical moment (photograph reproduced by Jacobs, 2014). In it, you see Glenn Burke, wearing his warmup jacket, sprinting from the Dodgers' third base dugout toward Dusty Baker. The pair's arms are raised with their palms facing one another in apparent preparation for the upcoming slap. The cause for jubilation on this occasion was actually Baker's grand slam home run off Philadelphia Philly's Jim Lonborg in the second game of the National League Championship series, which took place just three days later on October 5 (Shirley, 1977). How can we be certain of this fact? If you enhance the photograph and look very closely at the scoreboard in the background, then you'll see that this is the *fourth* inning of the October 5 game, not the *sixth* inning of the October 2 game when the famous Baker homer was hit! Bill Shirley's *Los Angeles Times* report of the October 5 game published the next day plus Andy Hayt's accompanying photograph provide definitive confirmation. This must therefore have been the *third* High Five between Burke and Baker.

BABY, WHAT A BIG SURPRISE

Returning to the memorable afternoon of October 2, it is critical to note that this *first* High Five was not planned; it was entirely unscripted and unrehearsed. Dusty Baker recounted the event in a marvelous 2014 ESPN film, *The High Five,* as follows: "It was just something that we did. Sometimes you don't know why you do some of the things you do, especially when you are extremely happy.

You just respond to each other." Baker took special note of how Burke was moving toward him: "His hand was up in the air, and he was arching way back. So, I reached up and hit his hand. It seemed like the thing to do."

It is commonly said that "one good turn deserves another." So, when Burke followed Baker in cracking his own home run against Richard, Dusty happily returned the favor with the *second* High Five.

MORE THAN A FEELING

The Dodgers promptly embraced this novel greeting, not permitting the excitement sparked by the High Five to fade like a fleeting feeling. They made it a signature celebratory ritual for the team. Vin Scully, the legendary radio and television announcer for both the Brooklyn and Los Angeles Dodgers, described the High Five in this way: "They are now doing what has become a Dodger trademark: they are doing a High Five. You don't shake hands straight on. You have to hold your arm fully extended over your head. Leap as high as you can. And, give up skin."

The Dodgers franchise also promoted the High Five in its fan merchandise and published materials. For the 1980 season, they created a special logo that was printed on t-shirts and, in 1981, they featured the High Five on the cover of the Dodgers yearbook. For the as-yet uninitiated, the Dodgers explained in their press guide that the High Five is customarily given after a home run, a good defensive play, or a Los Angeles victory.

Years later, there would be precious few uninitiated. The High Five had spread worldwide, permitting joy to be overtly expressed and shared with friends and teammates. Abdul-Jalil al Hakim, celebrity agent and Burke's boyhood friend, articulated the social significance of the High Five the following way:

> Look at where the High Five has taken society: from that gesture between him and Dusty to today. When you are caught up in that moment – that instant gratification, that exhilaration, that

exaltation, that achievement, and you want to be able to say "hey, look at what I've done" – the High Five started that. And, you can go anywhere around the world and people know what that is. They know what that means. Everyone wants that moment. And, the High Five made that accessible to everybody. When you look at how it's changed the world – and it's a universal symbol for all of us to share – there's no words you can attach to it. That's why there's no words necessary for it. It's a gesture that strikes your soul.

ONE IS THE LONELIEST NUMBER

For Glenn Burke to be so indelibly associated with such a unique expression of happiness, the remainder of his personal saga is utterly heartbreaking. The home run that Burke hit on October 2, 1977, was lamentably to be his *only* one for the Los Angeles Dodgers. On May 17, 1978, the Dodgers shockingly and unexpectedly traded Burke to the Oakland Athletics for another outfielder, Bill North.

This trade meant that Burke was forced to move from a *first-place* National League team with a 1977 record of 98 wins and 64 losses to a *last-place* American League team with a 1977 record of 63 wins and 98 losses. The trade did allow Burke to return to his home town, where he had been duly fêted as a star high school basketball player before the Dodgers drafted him for big league baseball. But, his fervent dream of playing out his career on a highly competitive major league team had been shattered. So too had his *joie de vivre*, which had made him such a favorite with his teammates, who considered "Burkey" to be the heart and soul of the ballclub. How could the Dodgers have so cruelly discarded Burke?

After all, the Dodgers franchise had famously signed Jackie Robinson, breaking the color barrier in Major League Baseball. A true baseball innovator, Brooklyn Dodgers President and General Manager Branch Rickey signed Robinson to a minor league contract in 1945 and promoted him to the major leagues in 1947.

Rickey was firmly determined to bring racial equity to the sport, having been widely quoted as saying: "I may not be able to do something about racism in every field, but I can sure do something about it in baseball." Robinson received further backing from Dodgers field manager Leo Durocher, who is frequently cited as firing back at his many critics: "I do not care if the guy is yellow or black, or if he has stripes like a ... zebra. I'm the manager of this team, and I say he plays."

Thirty years later, no such support was given to Burke when the Los Angeles Dodgers management learned of his homosexuality. Dodgers Vice President and General Manager Al Campanis – infamous for his overtly racist comments about blacks in sports management – gave little comfort to Burke, although he egregiously offered a honeymoon payment of $75,000 if Burke's marriage to a woman could be arranged. Burke scornfully refused Campanis' "helpful gesture" (Smith, 2013). Perhaps even more damaging to Burke's future in Los Angeles was his close friendship with the son of Dodgers field manager Tommy Lasorda. Tommy Lasorda, Jr. ("Spunky") was widely rumored to have been gay and died in 1991 at the age of 32 from complications arising from AIDS. Shipping Burke to the Athletics effectively cut short any serious relationship developing between him and Spunky.

So, off Burke went to Oakland, where his career sputtered due to multiple injuries and because of the offensive treatment he received from homophobic field manager Billy Martin (Frey, 1994). Burke put it this way:

> It got to the point where prejudice just won out. The Dodgers got rid of me, and everybody on the team knew why. Billy Martin didn't want no part of me, and no one else would sign me. I just got blackballed. A gay man in baseball? Uh, uh. No way.

The remainder of Burke's long downward spiral is detailed in his poignant autobiography, *Out at Home*, co-authored by Erik Sherman

(Burke and Sherman, 1995). Burke hoped that people would have the courage to read the book: "Not because of the money. After all, there's a chance I won't be around when this book goes public and won't see any of the money anyway." His remarks were sadly prophetic; he died of complications due to AIDS on May 30, 1995, at the age of 42, just before the book was to be published.

STILL THE ONE

Except perhaps for the participants in original sin, claims of original ownership are often disputed. This is also true for the High Five. Many stories swirl about its origin (Banks, 2016; Crockett, 2014; Mooallem, 2011). One goes as far back as ancient Egypt's Great Pyramid of Giza, where a painting portrays two people facing one another with raised palms. Then, there's a brief open-palm slap as two friends part in the 1960 French New Wave film *Breathless*. Yet another origin story points to the prior popularity of "giving some skin" or the "low five" among jazz entertainers. A more recent story credits the gesture to college basketball teammates who competed a few years after the 1977 Dodger Stadium event.

However, none of these stories provides the contemporary evidence or narrative consistency as clearly or persuasively as does the Burke–Baker episode; this account is still the one accepted by most authors.

THEY CAN'T TAKE THAT AWAY FROM ME

At the end of Jon Mooallem's excellent 2011 ESPN article, he relates the story of a newspaper reporter interviewing Glenn Burke long after his baseball career had ended. Reflecting on the High Five, Burke shared his perspective: "You think about the feeling you get when you give someone the High Five. I had that feeling before everybody else." Others may suggest that Glenn Burke was not the originator of the High Five. But, I'll continue to do so and to be heartened that Burke held steadfastly to that belief despite the many hardships he subsequently endured. No, no they can't take that away from him!

Song Titles

For readers who are interested, here are the songs that I chose for the title and section headings for this story:

"One! Singular Sensation." *A Chorus Line*. www.youtube.com/watch?v=PJR4E-p_QqI

"One Fine Day." The Chiffons. www.youtube.com/watch?v=CDGz4aoYxT4

"Celebrate Good Times . . . Come on." Kool & the Gang. www.youtube.com/watch?v=fA0f7lkefww

"Baby, What a Big Surprise." Chicago. www.youtube.com/watch?v=w0xcr93xx3A

"More Than a Feeling." Boston. www.youtube.com/watch?v=oR4uKcvQbGQ

"One Is the Loneliest Number." Three Dog Night. www.youtube.com/watch?v=HNjEPHvDxZQ

"Still the One." Orleans. www.youtube.com/watch?v=SdfW_2frXnE

"They Can't Take That Away from Me." Irving Berlin. www.youtube.com/watch?v=Wd7CZybUZWw

REFERENCES

Banks, A. (2016, July 5). How the Los Angeles Dodgers Birthed the "High-Five." *The Hundreds*. https://thehundreds.com/blogs/content/los-angeles-dodgers-high-five-glenn-burke

Burke, G. and Sherman, E. (1995). *Out at Home: The True Story of Glenn Burke, Baseball's First Openly Gay Player*. New York: Berkley Books.

Crockett, Z. (2014, July 23). The Inventor of the High Five. *Priceonomics*. https://priceonomics.com/the-inventor-of-the-high-five/

Frey, J. (1994, October 18). Once a Promising Ballplayer, Glenn Burke Is Dying of AIDS. *The New York Times*. https://archive.nytimes.com/www.nytimes.com/library/sports/baseball/090699bbo-bean-burke.html

Jacobs, M. (2014). *The High Five*. ESPN 30-for-30. Grantland feature film. http://grantland.com/features/30-for-30-shorts-high-five-invention-glenn-burke/

Mooallem, J. (2011, August 8). History of the High Five. *ESPN The Magazine*. www.espn.com/espn/story/_/page/Mag15historyofthehighfive/who-invented-high-five

Newhan, R. (1977, October 3). The Gang of Four. *Los Angeles Times*. https://search.proquest.com/docview/158375442/7E3E1A3335F48C9PQ/68?accountid=14663

Shirley, B. (1977, October 6). Baker Gets Hit of Year and Dodgers Are Even. *Los Angeles Times*. https://search.proquest.com/docview/158309744/fulltextPDF/E69F864ABF7F4B26PQ/1?accountid=14663

Smith, M. J. (2013, May 8). The Double Life of a Gay Dodger. *Deadspin*. https://thestacks.deadspin.com/the-double-life-of-a-gay-dodger-493697377

FURTHER MATERIAL

Glenn Burke Cut Down by Hypocrisy, AIDS. (1995, June 1). *SFGate*. www.sfgate.com/sports/article/Glenn-Burke-cut-down-by-hypocrisy-AIDS-3144951.php

PART II **Medicine**

6 The Apgar Score
"Millions Have Been Saved"

> I think there's no question that millions have been saved simply by the need to assess the baby at birth. Virginia Apgar legendarily came up with the score.
>
> Richard M. Smiley, 2018

Richard M. Smiley paid the above unique and well-deserved homage to the woman whose title he now holds as the Virginia Apgar M.D. Professor of Anesthesiology at Columbia University Medical Center, and chief of obstetric anesthesia at New York-Presbyterian/Columbia University Medical Center. You may not be familiar with Dr. Apgar's name; however, if you were born after 1953, then she may have been responsible for devising your very first test – administered 1 minute (and again 5 minutes) after you were delivered.

If you passed her test, then you might not take particular notice of Apgar's historic contribution to medicine. However, if you failed her test, then you might want to credit the good doctor with what may very well have been the lifesaving care you received – precisely because of your low score on her test.

According to Smiley (2018), before Apgar's scoring system rose to prominence, the label "stillborn" was routinely applied to newborn babies who were blue or whose breathing was weak or failing. Physicians had no accepted procedure for intervention or resuscitation; they believed those infants would not survive. Shockingly, those babies were left completely unattended, possibly to die – the sad fate that may have befallen my mother's baby brother. Said Smiley, "Before [the Apgar Score], you took the baby out, cleaned it, and hoped it lived. A large number of neonates could have survived if they had simply been given oxygen or warmed up."

The Apgar Score is currently used worldwide to assess newborns' physical condition. As one physician has noted, "Every baby born in a modern hospital anywhere in the world is looked at first through the eyes of Virginia Apgar" (Yount, 2008, 10).

Apgar developed the elegant scoring system in the early 1950s when she was then a Professor of Anesthesiology at Columbia University College of Physicians and Surgeons and Director of Obstetric Anesthesia at New York-Presbyterian Hospital. The Apgar Score is believed to have transformed the birthing process, turning the attention of the attending medical staff to the newborn as well as to the mother – the earlier focus of care.

THE APGAR SCORE IS INTRODUCED

Apgar published her most important scientific paper in 1953 (it was reprinted in 2015 in a special commemorative issue of *Anesthesia & Analgesia*). She began her report by stating its main aim: to establish a "simple, clear classification or 'grading' of newborn infants which can be used as a basis for discussion and comparison of the results of obstetric practices, types of maternal pain relief and the effects of resuscitation" (Apgar 2015, 1056). At that time, no systematic and easily implemented system could be relied on to judge the infant's condition. So, Apgar outlined the logic of her scoring method as follows,

> A list was made of all the objective signs which pertained in any way to the condition of the infant at birth. Of these, five signs which could be determined easily and without interfering with the care of the infant were considered useful. A rating of zero, one or two, was given to each sign depending on whether it was absent or present. A score of ten indicated a baby in the best possible condition. The time for judging the five objective signs was varied until the most practicable and useful time was found. This is sixty seconds after the complete birth of the baby. (1056)

The scientific core of Apgar's paper details the scoring scheme itself and offers extensive empirical support for its validity and utility. Type of delivery and method of maternal anesthesia were two main factors that attested to the merits of her grading system. Apgar found that infants' scores were higher after spontaneous delivery than after breech delivery or caesarian section, and they were higher after spinal anesthesia than after general anesthesia. Apgar concluded her paper with these succinct summary statements,

> A practical method of evaluation of the condition of the newborn infant one minute after birth has been described. A rating of ten points described the best possible condition with two points each given for respiratory effort, reflex irritability, muscle tone, heart rate and color. Various applications of this method are presented. (1059)

A SHORT STORY WITH LONG-LASTING CONSEQUENCES

It is common practice in the scientific literature to omit any narrative describing how the reported work came to be performed. Such revealing personal accounts are largely left to others – usually historians.

We are extremely fortunate to have available the engaging story of how the Apgar Score came to be, thanks to the scholarly work of Dr. Selma Harrison Calmes (2015). Calmes herself is a retired physician who took Apgar's written advice to female medical students and specialized in the field of anesthesiology. Beyond her own prominent medical career, Calmes has written several biographical articles about Virginia Apgar and other distinguished women in anesthesiology. In 1982, she co-founded the Anesthesia History Association.

Now, to the story. In the early 1950s, Apgar was instructing Columbia University medical students in anesthesia. She herself had become especially interested in the largely neglected field of obstetric anesthesia: the medical specialty historically devoted to the anesthetic care of women prior to, during, and following childbirth. An additional question of obvious medical importance was how various types of maternal anesthesia might affect the health and wellbeing of the newborn; but, it had not been systematically studied.

During breakfast one morning in the hospital cafeteria, one of Apgar's medical students commented on the necessity of evaluating the condition of newborns as well as their mothers. Sensing the significance of this flagrant unmet need, Apgar actively seized the opportunity,

> "That's easy, you'd do it like this." She grabbed the nearest piece of paper, a little card that said, in essence, "Please bus your own tray," ... and scribbled down the 5 points of the Apgar Score. She then dashed off to the labor and delivery suite to try it out. (Calmes, 2015, 1062)

Try it out, she did! Together with a research nurse, Rita Ruane, Virginia Apgar methodically tested and refined the score (see her demonstrating the administration of the Apgar Score, The National Foundation-March of Dimes, 1959) over seven and a half months at the Sloane Hospital for Women, beginning with a total sample of 2,096 babies and ultimately detailing the results of 1,021 in her 1953 paper! Apgar first presented her test and findings a year earlier at the International Anesthesia Research Society meeting in Virginia Beach.

Calmes points out some initial concerns and limitations of the scoring protocol. One matter revolved around who would actually do the scoring. Apgar herself believed that an impartial observer should score the newborn, worrying that those who delivered the baby might be "subconsciously" inclined to overestimate the quality of their own work and thereby inflate their assigned scores. Today, many different medical practitioners are carefully trained to administer the test: obstetricians, nurses, midwives, and emergency medical technicians.

Apgar had originally planned to score the newborns only 1 minute after birth to quickly identify those babies in greatest need of urgent medical attention. Still other medical workers scored the baby 5 minutes after birth to allow them sufficient time to assess the effectiveness of resuscitation, if needed. After a few years, it became standard practice to record each baby's 1- and 5-minute scores.

Finally, a sturdy metal clipboard containing both a paper scoring sheet and an attached timer – which rang a bell at 1 minute and 5 minutes after the baby's delivery – was developed by Puerto Rican anesthesiologist Colon-Morales (1971). This simple apparatus greatly standardized the timing and the accuracy of the recorded scores.

Although unrelated to the validity or reliability of the Apgar Score, its broad popularity may also have been importantly related to its name. Beyond bearing the name of its creator, "APGAR" is also a *backronym* which stands for the biological metrics the Apgar Score assesses: Appearance, Pulse, Grimace, Activity, and Respiration. This clever label was introduced in 1962 as a mnemonic device to help medical staff recall the various points of the score. Not only did this label stick but Dr. Apgar is said to have been amused whenever she encountered doctors in training who were stunned to learn that Apgar was actually a real person!

It is of historical note that the need for a catchy and memorable name was obvious to the editor of Apgar's original 1953 paper. Dr. Howard Dittrick was quite enthusiastic about her score and its potential clinical impact, "However, it needs a better name to help us remember it. If such a name can be found ... then there is a bright future for Apgar's score" (Wong and Shafer, 2015). The need was indeed met, albeit a decade after Apgar's article was published and by a person whose identity remains unknown.

VIRGINIA APGAR: A LIFE OF RESILIENCE AND EXCELLENCE

"Apgar's life story shows how history can shape lives." Thus, wrote Calmes (2015, 1063) before summarizing the influences that shaped Apgar's fascinating life as well as the profound effect that Apgar's work had on the history of medicine.

Virginia Apgar was born in Westfield, New Jersey, on June 7, 1909, the youngest of three children. Her mother, Helen, was the daughter of a Methodist minister. Her father, Charles, was an insurance executive as well as an amateur inventor and astronomer, Virginia often assisting him in his makeshift basement laboratory.

The Apgars were an especially active and highly musical family; Virginia learned to play the violin and cello as a child, and she kept playing as an adult. Virginia had many other interests, especially stamp collecting and sports.

At Westfield High School, Virginia excelled in her science classes, but she did not shine in her home economics classes – her friends declaring that she never learned how to cook! By the time she graduated in 1925, Virginia had chosen to pursue a career in medicine, a decision that may have been related to her eldest brother's death from tuberculosis and another brother's chronic childhood illness.

That same year, Apgar enrolled in Mount Holyoke College – a prestigious all-women's liberal arts school located in South Hadley, MA. Because her family was not wealthy, Apgar's studies had to be financially supported by several scholarships and part-time jobs. Apgar majored in zoology with minors in physiology and chemistry. By all accounts, she not only excelled academically but she made the most of many other collegiate experiences. In one of her letters home to her parents, Apgar confided: "I'm very well and happy but I haven't one minute even to breathe." (*The Life and Legacy of Virginia Apgar '29*, 2014). And, according to her Biographical Overview in the U.S. National Library of Medicine, the phrase "How does she do it?" became Apgar's trademark in college. To wit,

> She played on seven sports teams, reported for the college newspaper, acted in dramatic productions, and played violin in the orchestra. Even with all these activities, her academic work was exceptional; in her last year, her zoology professor and advisor [Professor Christianna Smith] noted, "It is seldom that one finds a student so thoroughly immersed in her subject and with such a wide knowledge of it."

Soon after graduation from Mount Holyoake, Apgar began medical school at Columbia University's College of Physicians and Surgeons – a mere 1 month before the calamitous Wall Street crash of October, 1929. Apgar was one of only nine women in a matriculating class of ninety. Despite continuing to experience financial difficulties, Apgar received her M.D. four years later. She graduated fourth in her class and was made a member of the medical honorary society Alpha Omega Alpha (*Changing the Face of Medicine. Dr. Virginia Apgar, n.d.*).

In 1933, Apgar began a 2-year internship in her preferred medical specialty, surgery, at Presbyterian Hospital (currently New York-Presbyterian Hospital/Columbia University Medical Center). Although she was an outstanding surgery intern, the smooth sailing of Apgar's professional development now encountered stormy seas. Her two chief troubles: the economic malaise of the Great Depression and the harsh reality of sexism.

Before completing her first year, Apgar's mentor, Dr. Alan O. Whipple, warned her that the upcoming years looked to be extremely detrimental to the professional advancement and financial security of female surgeons; he therefore advised Apgar to consider changing course. Whipple believed that, if surgery were to continue to advance, then innovations and improvements would be necessary in anesthesia – at the time, a discipline managed primarily by nurses and nonspecialist physicians. In Apgar, Whipple astutely sensed "the energy, intelligence, and ability needed to make significant contributions in this area."

Apgar decided to follow Whipple's sage advice. However, because anesthesiology was not to be accepted as a medical specialty for another decade, Apgar had considerable difficulty finding a suitable training program after completing her surgical internship in 1935. So, for a year, Apgar trained in Presbyterian's own nurse-anesthetist program.

Finally, although Apgar was unable to secure a longer-term appointment, she spent six months training with Dr. Ralph M. Waters as a "visitor" in the Department of Anesthesia at the University of Wisconsin-Madison. Waters is regarded as the father of academic anesthesiology, with the Wisconsin department being the first in the United States. Apgar found her training experience exceedingly valuable, but not especially hospitable. There, "she faced the usual woman physician problem, lack of housing for trainees, and had to sleep in Waters' office for 2 weeks until a room was found for her, in the maids' quarters. [Apgar also bristled] at having to miss department events held in male-only dinner clubs" (Calmes, 2015, 1061).

After finishing her brief stay in Wisconsin, Apgar returned to New York City and trained for six months in Bellevue Hospital with Dr. Emery A. Rovenstine, who had also studied with Waters. Rovenstine is best known for establishing the first academic Department of Anesthesiology at Bellevue.

Apgar returned to Presbyterian Hospital in 1938, now as Director of the newly created Division of Anesthesia within the Department of Surgery. This was a groundbreaking appointment, as it made Apgar the first woman to head a division at Presbyterian Hospital. In this position, Apgar assumed a variety of administrative responsibilities: recruiting and training anesthesiology residents and faculty; coordinating anesthesia services and research; and, teaching rotating medical students. Over a period of eleven years, Apgar took several steps to develop the fledgling program into a prominent division of Presbyterian; however, it proved difficult for her to recruit male faculty to work under the leadership of a female superior.

So, in 1949, Emanuel M. Papper, also a student of Rovenstine, was hired as a Professor and appointed as the Chair of the Division of Anesthesiology (and, in 1952, as the Chair of the newly created Department of Anesthesiology), effectively superseding Apgar as leader of Anesthesiology. Calmes (2015) suggests that Apgar considered Rovenstine's appointments to be unexpected and discouraging. This moment was to represent a "pivotal point" in Apgar's career.

Notwithstanding this disappointment, there was also reason for being optimistic: Apgar had been promoted, making her the first female full professor in the Columbia University College of Physicians and Surgeons. Apgar saw this as a timely occasion to take a duly-earned 1-year sabbatical. And, on her return to Columbia, she moved wholeheartedly into obstetric anesthesia, now conducting much of her clinical and research work at the affiliated Sloane Hospital for Women. Apgar's growing interest in the effects of maternal anesthesia on the newborn set the stage for that "breakfast breakthrough" and for her systematic development and refinement of the Apgar Score.

APGAR BEYOND THE SCORE

In the process of conducting research on her innovative scoring system, Apgar had attended over 17,000 deliveries during the 1950s. She had, of course, witnessed many regrettable cases of birth defects; those newborn malformities led Apgar to wonder whether they correlated with the scores she recorded from her test. In the hope of deploying more advanced statistical methods to better appreciate this possible relationship, Apgar took a sabbatical leave in 1958 and enrolled in the Master of Public Health program at the Johns Hopkins School of Public Health. Apgar swiftly obtained her master's degree in 1959.

Although planning to return fulltime to Columbia, another unexpected event changed the trajectory of Apgar's career. The National Foundation for Infantile Paralysis, better known today as the March of Dimes, was expanding its portfolio to incorporate childhood disabilities other than polio as well as to investigate the negative consequences of premature birth. Apgar was asked to head its new Division of Congenital Malformations and she accepted. In this capacity, Apgar was an exceptionally vigorous ambassador and advocate: traveling extensively, speaking effectively to diverse audiences, and helping to greatly increase funding for the foundation. Apgar also served the foundation as Director of Basic Medical Research (1967–1968) and Vice-President for Medical Affairs (1971–1974).

During her tenure at the March of Dimes, Apgar was concurrently active in several other capacities. She remained on the Anesthesiology faculty of the Columbia University College of Physicians and Surgeons until at least 1967, when she was lightly lampooned in the school's *Aesculapian* Yearbook declaring that an Apgar Score of 11 would simply be "Impossible!" She was also a lecturer (1965–1971) and a clinical professor (1971–1974) of pediatrics at Cornell University School of Medicine in New York City; there, she taught teratology (the study of birth defects), being the first to hold a faculty position in this new scientific area. Apgar was also appointed as a lecturer (1973) in medical genetics at the Johns Hopkins School of Public Health.

Throughout her long and distinguished career, Virginia Apgar received many awards. Among them, the U.S. National Library of Medicine lists honorary doctoral degrees from the Woman's Medical College of Pennsylvania (1964) and her alma mater Mount Holyoke College (1965), the Elizabeth Blackwell Award from the American Medical Women's Association (1966), the Distinguished Service Award from the American Society of Anesthesiologists (1966), the Alumni Gold Medal for Distinguished Achievement from the Columbia University College of Physicians and Surgeons (1973), and of particular importance and personal relevance, the Ralph M. Waters Award from the American Society of Anesthesiologists (1973). In 1973, Apgar was also elected Woman of the Year in Science by the *Ladies' Home Journal*.

Apgar exemplified a dedicated medical practitioner and scientist. She dealt with obstacles and disappointments by persisting and exploiting the opportunities that were afforded her. In so doing, Apgar achieved exceptional success in new frontiers of medicine, founding the field of neonatology, the area of medicine that specializes in premature and ill newborns.

Apgar's trailblazing career most assuredly advanced the cause of gender equality – by example and without fanfare. She privately expressed frustration with gender discrimination, but she publicly pronounced that "women are liberated from the time they leave the womb."

Although well known for her speedy lifestyle – she spoke fast, walked fast, and drove (much too) fast – liver disease slowed Apgar's pace during her later years. Nevertheless, she never retired. Virginia Apgar passed away on August 7, 1974, at Columbia-Presbyterian Medical Center, where she began her medical training and where she spent so much of her subsequent professional career.

Apgar also received two posthumous commendations. In 1995, she was inducted into the National Women's Hall of Fame. And, in 1994, she was honored with a commemorative US postage stamp as part of the Great Americans series. The 20-cent stamp bore Apgar's portrait and the words: "Virginia Apgar-Physician." As a stamp collector herself, I'm sure that this recognition would have been especially satisfying.

Indeed, for readers who might also be philatelists, the official First Day of Issue envelope containing Apgar's commemorative stamp bore additional lines of significance, "The APGAR Score for Newborns" and "Doctor Apgar stressed immediate attention to newborns."

Are these words too paltry to capture the full significance of Virginia Apgar's life and work? Let the closing words instead be those of Apgar's admiring biographer, Selma Calmes (2015),

> Her life was shaped by the history of women in medicine, the economic history of the early 1900s, the history of anesthesiology, and the history of polio. Virginia Apgar was an unforgettable character unique in American medicine. Her zest for life, for people, for the specialty of anesthesia, for newborns, for scientific research, and for her dearly loved music left an indelible memory on anyone she met. She also left us a lasting tool, the Apgar Score, providing a structured approach to evaluate newborns. Her score serves as a common language among the various specialties, including anesthesiology, that care for newborns. Her score led to better treatment of newborns and to great advances in anesthesia for their mothers. Her score was a unique contribution to anesthesiology, to maternal and child health, and to a generation of researchers dedicated to improved neonatal outcomes. (1063–1064)

REFERENCES

Apgar, V. (2015). A Proposal for a New Method of Evaluation of the Newborn Infant. *Anesthesia & Analgesia, 120,* 1056–1059 (Original work published 1953).

Baskett, T. (2019). Apgar, Virginia (1909–1974): Apgar Score. In *Eponyms and Names in Obstetrics and Gynaecology* (5–6). Cambridge: Cambridge University Press.

Calmes, S. H. (2015). Dr. Virginia Apgar and the Apgar Score: How the Apgar Score Came to Be. *Anesthesia & Analgesia, 120,* 1060–1064.

Changing the Face of Medicine. Dr. Virginia Apgar. (n.d.). National Institutes of Health. https://cfmedicine.nlm.nih.gov/physicians/biography_12.html

Colon-Morales, M. A. (1971). Apgar-Score Timer. *Anesthesia & Analgesia, 50,* 227.

The Life and Legacy of Virginia Apgar '29 (2014, July 14). Alumnae Association, Mount Holyoake College. https://alumnae.mtholyoke.edu/blog/the-life-and-legacy-of-virginia-apgar-29/

The National Foundation-March of Dimes (1959*). Virginia Apgar Demonstrating Apgar Score* [Photograph]. Retrieved October 19, 2020, from https://profiles.nlm.nih.gov/spotlight/cp/catalog/nlm:nlmuid-101584647X51-img

Smiley, R. M. (2018). Interviewed for: It Happened Here: The Apgar Score. Health Matters. New York-Presbyterian. https://healthmatters.nyp.org/apgar-score/ www.youtube.com/watch?v=dxz7qhIXBHc

Virginia Apgar. Biographical Overview (n.d.). *Profiles in Science.* U.S. National Library of Science. https://profiles.nlm.nih.gov/spotlight/cp/feature/biographical-overview

Wong, C. A. and Shafer, S. L. (2015). Dittrick's Missing Editorial about Apgar's Score. *Anesthesia & Analgesia, 120,* 962.

Yount, L. (2008). *A to Z of Women in Science and Math.* New York: Facts on File (Original Work published 1999).

7 The Ponseti Method
Effective Treatment for Clubfoot Only 2,400 Years in the Making

Clubfoot is a serious birth defect in which an infant's foot is twisted inward and downward (see a photograph by Brachet, 2005). It is usually an obvious and isolated defect for otherwise normal and healthy newborns. In babies with clubfoot, the tendons that connect the leg muscles to the heel are too short and the Achilles tendon is too tight. This congenital abnormality occurs in approximately one out of every thousand births. In about half of the babies affected, both feet are deformed. Boys are twice as likely as girls to have clubfoot. The cause of clubfoot is presently unknown, although it may have a hereditary basis. The disorder may be related to the mother's age or whether she smokes, takes drugs, or has diabetes.

If the condition is left untreated, then clubfoot does not spontaneously improve. As children develop, they may come to walk on their ankles or on the sides of their feet, creating a painful movement disability. Severe complications such as calluses and arthritis can also result if clubfoot is left untreated.

Many famous people have had clubfeet. Historical figures include Roman emperor Claudius, British Romantic poet and politician Lord Byron, and nineteenth-century American politician and abolitionist Thaddeus Stevens. Even athletic greats have been born with clubfeet, including two-time Olympic gold medal winning soccer player Mia Hamm, Super Bowl winning football player Charles Woodson, and two-time World Champion figure skater Kristi Yamaguchi. Such athletic success, of course, requires effective treatment.

The history of treating clubfoot (reviewed by Dobbs et al., 2000 and Sanzarello, Nanni, and Faldini, 2017) goes back to the famous Greek physician Hippocrates of Kos (460–377 BCE). Hippocrates is known as the father of modern medicine. Unlike earlier curative efforts founded on religious or magical beliefs, Hippocrates's therapies were based on meticulous scrutiny and rational reasoning (Yapijakis, 2009). He established the essentials of clinical medicine as it is practiced today: systematically examining the patient, assiduously observing the symptoms, rendering a logical diagnosis, and then fastidiously treating the patient. That empirical approach led Hippocrates to treat clubfoot with a method that has only recently been resurrected and elaborated by Ignacio Ponseti.

Just how did Hippocrates treat clubfoot? He used what is called manipulative correction: manually and progressively adjusting the foot toward its proper alignment and then applying sturdy bandages to hold the correction in place. Such manipulative therapy was best begun as soon after birth as possible and undertaken in a gentle and gradual manner, so that each successive adjustment was not too stressful to the bones and muscles of the developing infant. Hippocrates also recommended overcorrection, recognizing that some reversion was likely to happen as time passed and the infant grew. Finally, Hippocrates suggested that special shoes be worn after completion of the bandaging procedure, again to prevent clubfoot from relapsing.

Despite this most auspicious beginning, subsequent clubfoot treatment somehow failed to follow Hippocrates's innovative lead. In the ensuing centuries, a variety of different remedial techniques and devices were deployed to correct clubfoot. Pneumatic tourniquets, rigid braces, and other mechanical contrivances were used to twist the bones and muscles of the foot into their proper positions. These frequently forceful methods were of little therapeutic value, sometimes doing more harm than good.

Perhaps by default, surgery thereafter became the method of choice in treating clubfoot. Yet, despite the wide range of surgical techniques that were attempted, these too proved to be largely ineffective. For the 200,000 infants born each year with clubfeet, the prognosis remained poor – until Ponseti developed his technique. How did this valuable innovation come about?

Ignasi Ponseti Vives was born on June 3, 1914, in the town of Ciutadella on the small island of Menorca, which lies in the Mediterranean Sea off the east coast of Spain (see Luttikhuizen, 2011 for an excellent biography). Ponseti's father was an expert watchmaker, but a much less accomplished businessman. This reality led to many family relocations during Ponseti's boyhood as his father's business ventures faltered: first to two towns on the larger Mediterranean island of Mallorca and finally to Barcelona on the Spanish mainland. Working alongside his father, Ignacio acquired the intricate dexterities needed for watch repair – properly aligning the mainspring, the balance wheel and hairspring, the escapement, the gear train, the winding stem, and the dial train – manual skills that would later serve him well in dramatically different circumstances.

Ponseti proceeded to gain other valuable proficiencies during his five years at the University of Barcelona where, because of his exceptional school grades, he was granted tuition-free admission in 1931. Beyond excelling in purely academic coursework, Ponseti learned from his biology teacher how to dissect plant and animal tissues as well as how to slice and stain the brains of frogs for detailed microscopic analysis. Santiago Ramón y Cajal – the 1906 Nobel Prize winner in Physiology or Medicine and the recognized "father" of neuroscience – had taught at the same university from 1887 to 1892, where he developed the very same histological methods that Ponseti was later to master. These early biological studies thus set the stage for Ponseti's distinguished career in medical science and his many contributions to the pathology of skeletal growth disorders, most famously those involving the biomechanics of clubfoot.

Yet, whatever plans Ponseti may have had after taking his final medical school examinations on July 17, 1936, were radically altered the next day when the Spanish Civil War began. He soon found himself treating critically injured combatants, first in facilities in Spain and then in France from 1938 to 1939. Ponseti cured all sorts of war wounds with excellent results; indeed, even prior to the discovery of antibiotics, the available medical protocols he judiciously followed rarely led to infection. Ponseti also successfully treated fractures, first with traction and then with unpadded plaster bandages. In addition, he performed brain surgery as well as nerve sutures and tendon transfers; these especially delicate operations demanded prodigious manual adroitness.

With the turmoil continuing in Spain and his temporary residence in France threatened by imminent war, Ponseti sought a greater semblance of settlement. Thanks to Mexican President Lázaro Cárdenas's willingness to offer citizenship to all Spanish refugees, in July of 1939, Ponseti boarded the *Mexique* bound for Veracruz. Later journeying to Mexico City, he met an eminent orthopedist, Juan Faril, who himself had clubfeet. However, unable to find employment in the capital, Ponseti moved to Juchitepec, a small provincial town where he compassionately cared for its 5,000 inhabitants. Occasional visits to Mexico City sparked Ponseti's interest in studying orthopedic surgery with Dr. Arthur Steindler, Dr. Faril's mentor at The University of Iowa.

On the strength of his supportive letter of recommendation from Faril, Ponseti arrived in Iowa City on June 1, 1941, eager to begin his career as a medical scientist and practitioner. One of the first projects Steindler gave to Ponseti was to follow up twenty-four patients who had received clubfoot surgeries during the 1920s. Much to Steindler's disappointment, Ponseti reported that the patients' feet were stiff, weak, and painful. Radiographs revealed that most of the patients had misshapen bones and joints. Although Steindler believed that these poor outcomes might be overcome by improved surgical techniques, Ponseti was dubious.

Writing in 2007, Ignacio's urbane and erudite wife, Helena Percas-Ponseti, described this critical point in the clubfoot story,

> Ignacio was not convinced that [any new] clubfoot surgery, severing ligaments and joint capsules to align the bones of the foot, could show better results and not lead to stiffness, weakness, and pain. Surgery, no matter what improvements, would always be very

damaging to the foot. He was determined to find a noninvasive, safe alternative.

To this end, he studied many histological sections of aborted fetuses with clubfeet and also dissected clubfeet in stillborns to understand the pathology and the biomechanics of the deformed feet. Then he proceeded to devise a way to correct the deformity based on the functional anatomy of the tarsal joints and the gradual stretching of the contracted ligaments. In 1948 he started treating babies this way, and in 1963 he published [with Eugene Smoley a long-term] follow-up with his good results in the *Journal of Bone and Joint Surgery*. (21)

Percas-Ponseti then proceeded to answer the obvious question: Why was her husband able to correct the clubfoot deformity so effectively and to maintain that correction for such sustained periods of time?

Because the [embryonic cells] forming the skeleton in the baby's feet are mostly cartilaginous, soft, and easily molded into their right shape by manipulations following the functional anatomy of the feet. The joint surfaces reshape congruently after each manipulation. To maintain the correction, a plaster cast extending from the toe to the upper thigh is applied and worn for 4 or 5 days [Figure 7.1]. After 5 or 6 manipulations and castings, the clubfoot was corrected. To prevent a relapse, the child had to wear a [Denis Browne] bar with shoes attached at each end for three months and thereafter at night and napping hours for about three or four years [see Dolmanrg, 2008]. From then on, a relapse was extremely rare. Such treatment proved that the feet of these children develop as well as those of normal babies. (21)

Findings as phenomenal as these ought to have been exuberantly celebrated and widely disseminated. But, they were either dismissed as inconclusive or outright ignored!

Part of the problem in accepting Ponseti's method may have been that other physicians had in the early and mid-1900s tried

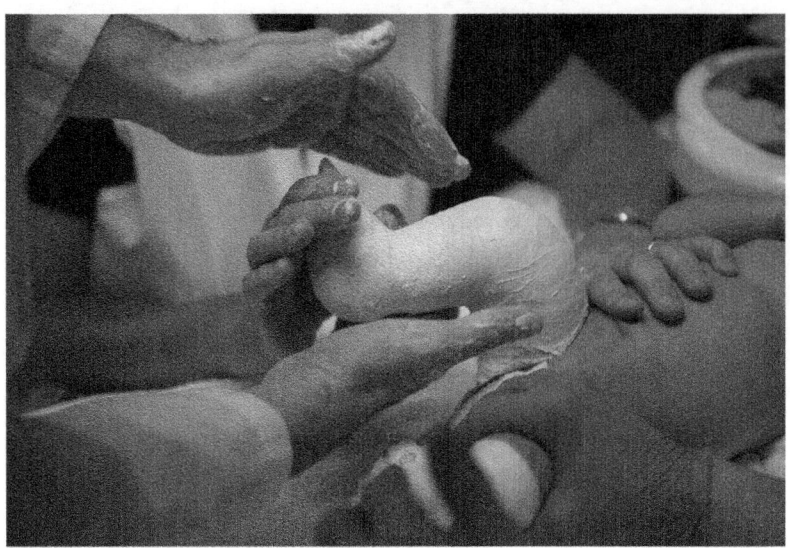

FIGURE 7.1 Dr. Jose Morcuende applying a plaster cast to the right foot of an infant.
Author: Tim Schoon. Courtesy of Tim Schoon, The University of Iowa

manipulative therapy with poor results. Why had they failed? "They just tried to smash the bones into position," Ponseti suggested.

Others, such as Michael Hoke and Joseph Kite, recognized that aggressive manipulation might actually damage the child's foot. They tried a gentler approach, but they too were largely unsuccessful. Ponseti believed that his predecessors simply "did not know how the joints moved." They had not done the necessary anatomical research that he had so painstakingly conducted. Nor had they so assiduously subjected the various elements of treatment to trial-and-error experimentation as he had. Finally, these physicians may have lacked Ponseti's deft touch. "You have to be able to feel every one of the bones with your hands. It's a little bit like playing the piano" stressed the watchmaker's son (Figure 7.2).

Beyond the unsuccessful efforts of these other researchers, doctors in general can be quite conservative. They can also be overly skeptical of innovative methods that they believe violate the

THE PONSETI METHOD 83

FIGURE 7.2 Dr. Ignacio Ponseti manually demonstrating the articulation of bones in the foot.
Author: Tim Schoon. Courtesy of Tim Schoon, The University of Iowa

foundations of their own prior training. Perhaps more extensive follow-up work by Ponseti would be sufficient to overcome the field's conservatism and skepticism.

With that hope in mind and at the urging and support of his wife, Ponseti (1996) gathered together the results of more extensive follow-up research and included it in his volume *Congenital Clubfoot: Fundamentals of Treatment* published in 1996 by Oxford University Press. This volume did prompt small contingents of physicians to journey to Iowa City to learn Ponseti's method. Nevertheless, it still failed to gain a firm foothold in medical practice, leading Ponseti to quip in frustration that "surgeons love their little knives."

What proved to be the real gamechanger? It was the Internet! Through the Internet, the parents of children with clubfeet who were averse to surgery were able to seek out noninvasive treatments and to widely broadcast their success stories (Morcuende, Egbert, and Ponseti, 2003).

In 1998, a web page with information on clubfoot treatment using the Ponseti Method was posted in the Virtual Hospital of The University of Iowa. Shortly thereafter, there was a striking change in patient referral patterns. In the years prior to 1998, an average of only five patients per year had been seen in Ponseti's clinic, most of them from Iowa. This number rapidly rose to about sixty patients per year by 2001, with most of them coming from out of state. Some 75 percent of the patients attending the clinic were self-referrals. Clearly, news of the Ponseti Method and its successful implementation at Iowa had gone viral!

Given the groundswell of public support it was receiving, in August of 2006, the American Academy of Pediatrics at last came to endorse the Ponseti Method as the most successful noninvasive and cost-effective clubfoot treatment and urged its use worldwide. So too did the American Academy of Orthopedic Surgeons and more than fifty orthopedic societies around the globe.

The Ponseti International Association at The University of Iowa has, since 2006, been working to make the benefits of Ponseti's

Method available to every child born with clubfoot everywhere. It is headed by Dr. Jose A. Morcuende, Ponseti's accomplished Spanish successor.

It is one thing to have developed the Ponseti Method; however, it is quite another to teach others to use the technique. It initially proved difficult for doctors to understand the intricate biomechanics of the tarsal joints necessary to correct clubfoot. So, Ponseti worked with a local craftsman to fashion a clubfoot model. After several iterations, the final product was made of plastic bones and elastic strings to simulate ligaments. This model proved invaluable to demonstrate to physicians the proper remedial maneuverings. Several small, soft, plastic feet were also created to exemplify the successive stages of correction for teaching purposes.

What a story! From humble beginnings on a remote Balearic Island, a young peripatetic doctor arrives in Iowa speaking little English. Nevertheless, building on his rich personal and medical experiences, he devises today's gold standard in clubfoot treatment – one bearing a remarkable resemblance to that originally developed by Hippocrates over 2,000 years earlier. We are far from fathoming how Ponseti's many and varied life experiences came together, but crediting his achievement to "insight" or "genius" seems woefully superficial.

Fortunately, we do have some of Ponseti's own words (Wilcox, 2003) to help us comprehend his creative process. Here is how he discussed the results of his examining the feet of stillborn fetuses. "We know that the foot grows normally for the first several weeks in the pregnancy and then something happens that we don't understand yet, and the foot twists. Under the microscope, I examined the ligaments and saw that they are wavy. You can stretch them gently without any damage." Ponseti suspected that this plasticity might be exploited by an iterative process of stretching and casting, along the general lines ascribed to Hippocrates. Ponseti then set about testing that general notion with his own specific techniques.

That all sounds simple enough, but recall that Hoke and Kite had little success with this process alone. Proper alignment of the baby's tiny bones is absolutely essential for effective treatment. Fully understanding the anatomy of the foot and appreciating how the clubfoot disorder develops in the first place proved to be critical to Ponseti's success. With that understanding, "You invite the foot to come back in the normal position. Already, the foot knows what this position is. It was in that position for the first half of the pregnancy. The foot was normal until then." In short, Ponseti's key to curing clubfoot was to reverse the process. That made perfect sense to a man as familiar as Ponseti with "turning back the clock."

Ignacio Ponseti passed away on October 18, 2009, in Iowa City (Figure 7.3 shows him still working at the university hospital in his final years). His life-changing method endures.

FIGURE 7.3 Dr. Ignacio Ponseti viewing an X-ray of a patient's feet.
Author: Tim Schoon. Courtesy of Tim Schoon, The University of Iowa

REFERENCES

Bergerault, F., Fournier, J., and Bonnard, C. (2013). Idiopathic Congenital Clubfoot: Initial Treatment. *Orthopaedics & Traumatology: Surgery & Research, 995,* S150–S159.

Brachet, Y. (Photographer). (2005). Bilateral Clubfoot in an Infant with Both Feet Rotated Inward [Photograph]. Retrieved October 19, 2020, from https://commons.wikimedia.org/wiki/File:Pied_bot,_varus_%C3%A9quin_(bilateral).jpg

Dobbs, M. B., Morcuende, J. A., Gurnett, C. A., and Ponseti, I. V. (2000). Treatment of Idiopathic Clubfoot: An Historical Review. *Iowa Orthopaedic Journal, 20,* 59–64.

Dolmanrg (Photographer). (2008). The Denis Browne Bar for Maintaining Clubfoot Adjustments after Cast Removal [Photograph]. Retrieved October 19, 2020, from https://commons.wikimedia.org/wiki/File:Botas.JPG

Luttikhuizen, F. (2011). Professor Ignaci Ponseti i Vives (1914–2009). *Contributions to Science, 7,* 205–214.

Morcuende, J. A., Egbert, M., and Ponseti, I. V. (2003). The Effect of the Internet in the Treatment of Congenital Idiopathic Clubfoot. *Iowa Orthopaedic Journal, 23,* 83–86.

Percas-Ponseti, H. (2007). *Homage to Iowa: The Inside Story of Ignacio Ponseti.* Iowa City, IA: The University of Iowa Printing Department.

Ponseti, I. V. (1996). *Congenital Clubfoot: Fundamentals of Treatment.* New York: Oxford University Press.

Sanzarello, I., Nanni, M., and Faldini, C. (2017). The Clubfoot over the Centuries. *Journal of Pediatric Orthopaedics B, 26,* 143–151.

Wilcox, C. (2003, February). A Healing Touch. *Iowa Alumni Magazine.*

Yapijakis, C. (2009). Hippocrates of Kos, the Father of Clinical Medicine, and Asclepiades of Bithynia, the Father of Molecular Medicine. *In Vivo, 23,* 507–514.

FURTHER MATERIAL

Several of Ponseti's quotations are from www.latimes.com/nation/la-me-ignacio-ponseti27-2009oct27-story.html

Ponseti training https://vimeo.com/9546017

Ponseti orthopedic devices https://mdorthopaedics.easyordershop.com/easyorder/index

Unique biographical material about Ponseti www.galeriametges.cat/galeria-fitxa.php?icod=LMJ#googtrans(ca|en)

Nice feature on Ponseti https://magazine.foriowa.org/story.php?ed=true&storyid=1831

8 The Heimlich Maneuver

Tennessee Williams – one of America's most prolific and prominent twentieth-century playwrights (and a graduate of The University of Iowa, I should add) – tragically died in his two-room suite at the Hotel Elysée at 60 East 54th Street in New York City on February 25, 1983. The seventy-one-year-old author expired from asphyxiation after accidentally swallowing or inhaling a plastic over-cap used to cover small containers of nasal spray or eye wash.

According to the National Center for Biotechnology Information (Duckett, Velasquez, and Roten, 2020), death that results from choking on foreign objects, such as coins, small plastic parts, or toys, is more likely in very young children than in adults. On the other hand, older adults are most likely to choke on food. Indeed, just such an instance occurred in Queensland, Australia, and was conspicuously featured in the international press.

The unfortunate incident occurred in conjunction with a celebration of the country's national day on January 26, 2020. All across Australia, parades, fireworks, and food festivals are held annually to mark the arrival of the First Fleet of British ships to Sydney Harbor in 1788, thus beginning the European settlement of the continent and the subsequent founding of the country of Australia. The Beach House Hotel at Scarness near Hervey Bay in southeast Queensland was the site of one such festivity – a spirited Australia Day lamington speed-eating competition. A lamington is a small cube-shaped cake which is dunked into chocolate and sprinkled with dried coconut. The cake is said to have been created in Queensland in 1900 by the chef of its eighth governor, Lord Lamington (Wong, 2019).

The celebratory event turned tragic, however, when a sixty-year-old woman began to choke on the cake. "Witnesses say the pub's security guard and manager rushed to her aid and started performing CPR while an ambulance was called. Paramedics took over when they arrived and she was taken to Hervey Bay Hospital but could not be revived and later died" (Yeung, 2020).

Without second guessing those trying their best to save the woman's life under especially stressful conditions, one is struck by there being no mention in the press accounts of anyone attempting to use the Heimlich Maneuver. Also referred to as the abdominal thrust technique, the Heimlich Maneuver is a highly effective means of expelling foreign objects which may be blocking a person's airway. The procedure involves first wrapping your arms around a person from behind, balling up one hand into a fist and placing the other hand around it, gripping the person's abdomen above the bellybutton and under the rib cage, and then sharply thrusting inward to force a surge of air from the lungs to clear the object from the windpipe – in a decidedly bellows-like manner.

Since its popularization in 1974, the Heimlich Maneuver has been an invaluable first-aid procedure that is believed to have saved the lives of countless thousands of choking victims. It is recommended by the American Medical Association, the American Heart Association, the American Red Cross, and premier medical institutions across the country including Johns Hopkins Medicine. Even so, suffocation by ingestion or inhalation remains the fourth most common cause of preventable death in the United States according to the Centers for Disease Control and Prevention. Given these statistics and the unfortunate lamington-eating fatality, it would seem that Americans and Australians alike would do well to increase their awareness of the Heimlich Maneuver.

ORIGIN OF THE MANEUVER

Henry Judah Heimlich (1920–2016) was a prominent thoracic surgeon and health scientist. He received his medical degree in 1943 from Cornell University Medical College in New York City and was later affiliated with the Jewish and Deaconess Hospitals in Cincinnati. In

1977, he was appointed Professor of Advanced Clinical Sciences at Xavier University. Heimlich's work on the abdominal thrust garnered the Albert Lasker Public Service Award in 1984. Heimlich was later inducted into two halls of fame: Engineering and Science in 1985 and Safety and Health in 1993.

As a young medical student, Heimlich became captivated by saving lives (Heimlich, 2016). His first life-saving deed came in 1941 at the scene of a railway accident. On that occasion, Heimlich helped rescue Otto Klug, a locomotive fireman, from a truly life-threatening situation. Klug's leg was severely injured and pinned under the car of an overturned train. Making matters worse, Klug and the train wreckage were slowly sinking into a pond. Heimlich held the victim's head out of the water and helped reduce the pain and stress of Klug's perilous entrapment until he could finally be freed by other rescuers.

Press accounts of the heroic rescue were featured in *The New York Times* and *The New York Daily News*. The main lesson that Heimlich learned from that episode was that medical expertise and common sense could help save lives, all of which he found deeply satisfying. Of course, the public recognition he received was also rewarding as was the gold watch he was awarded from the Greater New York Safety Council "for his calmness and courage in saving a life in a railroad wreck new South Kent, Conn."

Heimlich's subsequent career in medical practice and science allowed him to save many more lives in a variety of different ways. However, his most famous innovation was the development of the abdominal thrust technique, popularly called the Heimlich Maneuver. Because he was never shy to promote his accomplishments, we are indebted to Heimlich for providing a first-hand account of his important discovery. As his historical essay for the American Broncho-Esophagological Association makes clear, it was neither a simple nor a straightforward process of discovery.

Heimlich begins this captivating narrative in 1972. A *New York Times Magazine* article caught his eye. It noted that choking was then the sixth leading cause of accidental deaths. As he had for twenty years been interested in people's swallowing problems and had himself developed an effective esophagus replacement operation, Heimlich wondered whether he might be able to reduce the incidence of deaths from choking.

Heimlich's first step was to conduct library research. He learned from medical journals and texts that, since 1933, the American Red Cross had recommended slapping choking victims on the back to disgorge food obstructions from the airway. Heimlich was taken aback by this advice. Not only did he discover that this recommendation lacked sound scientific support, but he believed that this maneuver might be downright dangerous or even deadly; rather than expelling the obstruction, such blows to the victim's back might actually push the object farther down and wedge it even more tightly into the windpipe.

What appeared to be needed was a way to use the lungs to the advantage of the patient. Heimlich reasoned in the following fashion: "I decided that since back slaps caused deaths by forcing the object back downward in the airway the answer lay in creating a flow of air upward out of the lungs, using the lungs like a pair of bellows. To work up enough force to expel the object, I would have to find a way to compress the lungs sufficiently to create a strong flow of air out of the mouth." Discovering just how to do so would prove to be a matter of considerable trial-and-error experimentation.

In 1973, Heimlich began an extensive series of experiments in his laboratory at Cincinnati's Jewish Hospital. The subjects in these investigations were four 38-pound anaesthetized beagles. Heimlich attached a small balloon to the lower end of an endotracheal tube (a device which is commonly deployed to assist breathing) and inserted the balloon into the dog's airway. When the balloon was inflated, it could obstruct the windpipe, just like a foreign object. Heimlich hypothesized: "If I could produce a large flow of air by compressing the lungs, the tube should move upward out of the airway." That, he suspected, would be the solution to the problem.

Nevertheless, Heimlich's first tests were not at all successful. Simply pressing on the dog's chest failed to move the tube. So, chest compression proved to be insufficient. Why? The dog's strong rib cage barely budged. After all, the ribs function well to protect the lungs from undue compression.

Heimlich then tried a different approach: "I considered if you pushed the diaphragm upward into the chest, you would markedly reduce the volume of the chest cavity, which would compress the lungs. I pressed my fist above the dog's belly button and just under the rib cage, thereby pushing the diaphragm upward into the chest. Instantly the tube shot out of the animal's mouth! Repeating this procedure, I found the same result every time."

Perhaps it was the case that this stunning success was specific to Heimlich's peculiar experimental contraption: the endotracheal tube and balloon. What about a more natural obstruction? To answer this question, Heimlich dispatched his assistant to the hospital cafeteria to retrieve a ball of raw hamburger. Both of the earlier approaches – chest compression and the abdominal thrust – were once again assessed in the interest of scientist rigor. "Putting the meat into the animal's larynx or trachea, I again pressed the chest repeatedly. Nothing happened. Then I pressed upward on the diaphragm. The meat shot out of the dog's mouth! I repeated the procedure over and over. Each time it worked."

Perhaps the maneuver was simply specific to beagles. No matter how much pet owners may love their canine companions, Heimlich's life-saving maneuver was meant to help people – not dogs. So, Heimlich and ten of his fellow hospital colleagues participated in a series of tests to measure the air flow from the mouth when the diaphragm was pushed upward with a fist. The results confirmed that the induced flow of air ought to be more than adequate to expel an obstructing object – in people.

Fine, the abdominal thrust worked well. But, what was the simplest and most effective way to perform it? This was a particularly important question to answer because it was going to be a layperson – not a trained medical professional – who was most likely to serve as the victim's rescuer. Heimlich considered many options,

> You could brace the victim's back against a wall and push with your fist against his or her upper abdomen. You could lay the victim on the ground and push with your hand or foot on the upper abdomen. However, after many tests and trials, it became obvious that the best technique was to stand behind the victim and reach around the choking person's [waist] with both arms. Make a fist. Place the thumb side of your fist below the rib cage, just above the belly button, grasp the fist with the other hand and press it ... inward and upward. Perform it firmly and smoothly and repeat until the choking object is dislodged from the airway. In most reports the object flies out of the mouth and sometimes hits the wall or even the ceiling. It can be mastered in one minute from a poster. To learn it, you do not even need to take a first-aid course.

With all of this experimentation behind him – plus his developing variations in the basic method that could accommodate very large or very small choking victims as well as victims who had fallen or lost consciousness – the final challenge for Heimlich was to publicize the maneuver. All of this effort would be worthless if no one knew of it. Here, Heimlich's liking of the limelight came in handy.

Heimlich called the editor of a magazine, *Emergency Medicine*, who was familiar with and receptive to his prior research. The editor agreed to publish Heimlich's paper – flamboyantly entitled *Pop Goes the Café Coronary* – as well as to alert a friendly reporter at the *Chicago Daily News*, Arthur Snider. The reporter prepared a syndicated column, which was published in several hundred newspapers on June 9, 1974.

Heimlich found this publicity was most welcome. But, despite the derring-do of Heimlich's arranged announcement, the Heimlich Maneuver had at that point *never* actually been used to save a person from choking! Nonetheless, a reliable report of the first person ever to be saved by his abdominal thrust soon came. This is how Heimlich related the episode in his own historical essay,

A week later on June 16, the following report appeared on the front page of the *Seattle Times*. The headline read: "News article helps prevent a choking death."

A Hood Canal woman is alive today because of an article in some Sunday editions of *The Seattle Times*. Isaac Piha said he read the article twice Saturday night while in his cabin on Hood Canal. It told of Dr. Heimlich's method of forcing a piece of food out of the windpipe.

Piha, a retired restaurateur, was interested because of the number of instances patrons have choked to death on pieces of meat. Piha and members of his family were enjoying a Father's Day gathering Sunday afternoon when Edward Bogachus ran from his nearby cabin calling for help for his wife, Irene. Piha ran to the Bogachus cabin and found Mrs. Bogachus slumped at the dinner table and beginning to turn blue. "I thought about heart attack and about that article in *The Times* while I was running to the cabin," Piha said. "When I saw that they'd been eating dinner, I knew it was food lodged in her throat."

Mr. Piha became the first person to perform what was soon known as the "Heimlich Maneuver." He dislodged a large piece of chicken from Irene Bogachus's throat and she quickly recovered.

Irene Bogachus thus became the first person to benefit from Heimlich's important discovery. Indeed, she lived another twenty-two years, passing away in 1996 at the age of 86. Although Bogachus was the first to be saved by the Heimlich Maneuver, she would not be the last. Among the thousands to have been saved by the maneuver, we can add President Ronald Reagan; actors Elizabeth Taylor, Goldie Hawn, Jack Lemmon, Walter Matthau, Cher, Carrie Fisher, Ellen Barkin, Nicole Kidman, and Halle Berry; basketball sportscaster Dick Vitale; and, former New York Mayor Ed Koch.

EPILOG

Given such a worthwhile medical contribution as the Heimlich Maneuver, we would expect nothing but plaudits to have come

Henry Heimlich's way. Unfortunately, several instances of purported misconduct arose in Heimlich's later years, which have somewhat tarnished his legacy. Especially embarrassing is the fact that many of these alleged transgressions were publicized by his son, Peter (Heimlich, P. M., 2019).

Reflecting back on the impact that Heimlich's first life-saving experience had on his later work and recognition, it is indeed ironic that one especially critical exposé (Francis, 2004) went so far as to proclaim that: "Heimlich always coveted fame. Today, he's notorious."

For our purposes, I believe that we should accord much greater weight to the persistence and resourcefulness of Heimlich's trailblazing efforts. They beautifully exemplify the power of trial-and-error investigation in yielding innovative solutions to vexing human problems. Never knowing just what was going to work, Heimlich was undaunted in testing numerous alternatives until success was ultimately attained. Whatever his motivations, the effectiveness of his maneuver now speaks for itself.

REFERENCES

Duckett, S. A., Velasquez, J., and Roten, R. A. (2020). *Choking*. National Center for Biotechnology Information Bookshelf. National Institutes of Health. www.ncbi.nlm.nih.gov/books/NBK499941/?report=printable

Francis, T. (2004). Playing Doctor. *Cleveland Scene,* October 27. www.clevescene.com/cleveland/playing-doctor/Content?oid=1488395

Heimlich, H. (n.d.). Historical Essay: The Heimlich Maneuver. American Broncho-Esophagological Association. www.abea.net/historical-essay-the-heimlich-maneuver

Heimlich, H. J. (2016). *Heimlich's Maneuvers: My Seventy Years of Lifesaving Innovation.* Amherst, NY: Prometheus.

Heimlich, P. M. (2019). *Outmaneuvered: How We Busted the Heimlich Medical Frauds.* http://medfraud.info/

Wong, J. (2019, January 23). Everything You Need to Know about Lamingtons, the Most Australian Cake. *ABC Life.* www.abc.net.au/life/everything-you-need-to-know-about-lamingtons-australian-cake/10720880

Yeung, J. (2020, January 26). A Woman Choked to Death Eating Australia's National Cake on the Country's National Day. *CNN*. www.cnn.com/2020/01/26/australia/australia-day-lamington-death-intl-hnk-scli/index.html

FURTHER MATERIAL

Johns Hopkins Medicine (n.d.). *Choking and the Heimlich Maneuver.* www.hopkinsmedicine.org/health/wellness-and-prevention/choking-and-the-heimlich-maneuver

Radel, C. (2016, December 17). Dr. Henry Heimlich Dies at 96. *Cincinnati Enquirer*. www.cincinnati.com/story/news/2016/12/17/cincy-native-dr-henry-heimlich-dies-96/95556716/

9 Eating to Live

The Lifesaving Contribution of Stanley Dudrick

Toward the end of his long, painful struggle with colon cancer, my dad shared a profound thought with me. Always an avid eater, he confided: "I used to live to eat. Now, I eat to live."

As a pleasure, we take eating as a hedonistic given. Endless television shows feature fabulous foods from far-flung lands and exotic cultures, all titillating our taste buds. As a necessity, we rarely give food a second thought, except perhaps when we lament the persistent problem of undernourishment, which stood at 10.8 percent of the population in 2018 (Food and Agriculture Organization of the United Nations, 2019).

But, for millions of unfortunate people, normal eating is impossible. Those may include persons struggling with chronic bowel disorders, individuals suffering from extensive burns, and patients recovering from major surgery. Not only adults but also newborns and young children are vulnerable to such misfortunes.

Stanley J. Dudrick (1935–2020) was not the first surgeon to confront this grave reality, but he was the first to devise an effective method to feed those who would surely perish from undernourishment. In a 2013 Misericordia University interview, Dudrick recalled a particularly trying time in 1961 when three of his surgical patients died: "When they died, a piece of me died. You could do great operations, but many of the patients still died." The young surgery resident at the University of Pennsylvania Hospital came to realize that the life supportive resources that were available for his patients simply did not measure up to the technical resources that were available for him to perform highly skillful surgeries. "If the patient doesn't have enough reserve in the bank, from a nutritional standpoint, they run out of fuel and substrate, and they cannot heal and cannot fight infection or restore their strength." What is the surgeon to do under these dire circumstances?

At the urging of his mentor, Dr. Jonathan E. Rhoades, Dudrick set out to develop a functional system for providing patients with Total Parenteral Nutrition (TPN): a means for delivering complete intravenous nutrition, in which liquid food is injected directly into the bloodstream via a catheter connected to a vein, thereby circumventing the stomach and small intestine. From the outset, Dudrick knew that this was not going to be an easy task.

According to Dudrick's (2003) detailed description of the early history of TPN: "The prevailing dogma among clinicians in the 1960s was that feeding a patient entirely by vein was impossible; even if it were possible, it would be impractical; and even if it were practical, it would be unaffordable. Indeed, TPN was considered a 'Gordian Knot' or a 'Holy Grail' pursuit by most physicians and surgeons" (292).

In order to succeed in this quest, an intimidating number of technical obstacles had to be considered and overcome. Which vein would prove to be the most practical for injecting food? How much fluid could patients safely tolerate on a daily basis? What exact combination of nutrient substances would be necessary to sustain the patient? How could those nutrients best be combined, stored, and distributed? These and dozens of other extraordinarily difficult questions remained unanswered. "Furthermore, it was essential to overcome decades of written and verbal expressions by prominent physicians and scientists that long-term TPN was impossible, improbable, impractical, or folly. Plausible fundamental evidence to the contrary had to be generated if skepticism and prejudices were to be neutralized or overcome and if widespread clinical acceptance were eventually to occur" (293).

Dudrick accepted this supreme challenge with intense fervor and fortitude: "Accordingly, efforts were directed toward designing experiments in the laboratory to explore and verify the efficacy and safety of TPN with the goal of applying clinically to patients the basic knowledge, skills, and techniques acquired, developed, and mastered in animals" (293).

Dudrick and his team began research with both adult and newborn mongrel dogs. Several months of tedious trial-and-error experiments were required to create a feeding formula, in which all of the necessary ingredients – proteins, carbohydrates, fats, vitamins, and minerals – could be properly blended and dispensed through the animals' venous system. These extensive preliminary projects set the stage for what was to be a key experiment, which studied more genetically comparable and consistently proportioned puppies.

From 1965 to 1966, a group of six male pedigreed Beagle puppies (see Figure 9.1) were fed entirely by central infusion through the superior vena cava (the large vein that returns blood from the head, neck, and upper limbs to the heart) for 72–256 days; these puppies were comprehensively compared with a group of six of their size- and weight-matched littermates that were fed orally. Dudrick excitedly reported that "The 6 puppies fed entirely IV outstripped their control orally fed littermates in weight gain and matched them in skeletal growth, development, and activity" (293). This proof-of-concept experiment proved to be a stunning success!

It was then time for the next and most important step – moving the innovative venous feeding protocol from dogs to human beings: "Having thus demonstrated beyond a doubt that it was possible and practical to feed animals entirely by vein for prolonged periods without excessive risks or compromises of growth and development potential, attention was directed toward applying what had been learned in the laboratory to the treatment of surgical patients" (296). In this study, six seriously malnourished patients suffering from chronic gastrointestinal maladies were catheterized and nourished entirely by vein for a period of 15–48 days. The life-sustaining formula they were given was, of course, suitably modified from the puppy recipe. Despite their originally grave prognoses because of severe malnourishment, all six of the patients were discharged from the hospital in good condition.

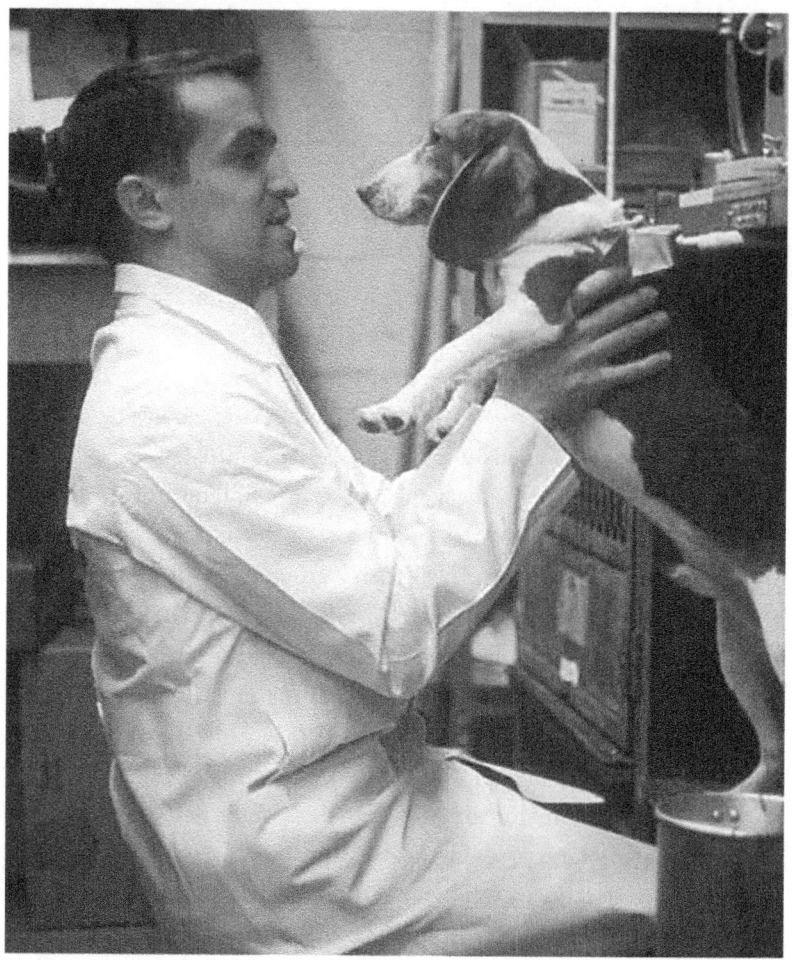

FIGURE 9.1 Dr. Stanley J. Dudrick with one of his beagles, Stinky. Courtesy of Theresa M. Dudrick

A few months following this groundbreaking human study, a unique challenge presented itself: might Dudrick's experience in intravenously feeding puppies prove applicable to human infants? In July of 1967, a baby girl in the Children's Hospital of Philadelphia had been operated on for a severe bowel obstruction. However, the traumatic surgery had left her slowly expiring from starvation. After extensive medical and ethical deliberations, the hospital staff decided that TPN represented the only reasonable means to save her life. Over a period of twenty-two months, the infant gained weight and grew in height and head circumference; indeed, she was normally active for her age. Unfortunately, her progress could not be sustained.

"Although she eventually died, clinicians gained a tremendous experience metabolically and technologically during her treatment, and her legacy to parenteral nutrition is unparalleled" (297) wrote Dudrick. Indeed, in his later Misericordia University interview (2013), Dudrick went even further. "We learned more from that child than all of the rest of the knowledge we've acquired since then in this field because we had such rudimentary resources to start with, and we had to innovate, hone, and try them on her, and improve them." Since then, many important refinements have been and are continuing to be made in order to allow children to develop into healthy and happy adults.

Nevertheless, as time has passed, some have occasionally questioned the ethics of keeping children alive if they must, even on a limited schedule, be tethered to an external feeding device. As late as 1983, concerns were still being raised about the sustainability of TPN (Kleiman, 1983). Dudrick's answer to these criticisms then is worth repeating now: "We're dealing with an evolving medical science. *As we go along, we modify things and learn* [emphasis added]. But to deny T.P.N. to a child unless there is absolutely no hope is simply unethical. It is letting a child starve to death" (Kleiman, 1983).

Currently, TPN is a truly life-sustaining and life-saving therapy, one which represents the culmination of Dudrick's passionate pursuit. The mortality rate in post-surgical cases has dropped from between 40 percent and 60 percent to only 5 percent, in the process saving well over 10 million lives (Smith, 2020). Further plaudits have come from collaborators, professionals, and patients who recently wrote a stirring tribute to Dudrick (Tappenden et al., 2020). Coalescing many of those accolades and remembrances provides a very human perspective of the man and his work,

Dr. Dudrick's impact on human life was immense. TPN provides a *literal lifeline* to millions of patients; it is now recognized as one of the three most significant achievements in modern surgery, together with asepsis and antibiotic therapy. From newborns to the aged, Dudrick's development of TPN meant the difference between certain slow death from malnutrition and a full, healthy life. Dudrick always believed that his greatest contribution was taking the sickest, most malnourished critically ill patients, and nutritionally supporting them to get them through an operation, to rehabilitation, and back home.

Dr. Dudrick was never shy about the values that shaped his determination to overcome obstacles and to develop the life-saving therapy of TPN. He often spoke of the courage shown by his grandfather, who had emigrated from Poland to work in the coal mines, and the work ethic of his father, who also worked as a coal miner when Dudrick was a boy. Later, Dudrick's father held other jobs, some requiring technical work constructing homes and repairing cars, and often involving the help of his handy son. Imagine how impactful these influences were in a small Pennsylvania coal mining town in the 1930s.

Additional biographical details now seem appropriate. Stanley John Dudrick was born on April 9, 1935, in Nanticoke, Pennsylvania. He was the eldest of four children. After high school, Dudrick graduated with honors from Franklin and Marshall College in Lancaster, Pennsylvania with a B.S. degree in biology in 1957; there, he was elected class president and was awarded the Williamson Medal as the outstanding student of his graduating class. Having been impressed by a doctor's having cured his mother's life-threatening fever, Dudrick pursued post-baccalaureate training at The University of Pennsylvania School of Medicine. There, Dudrick was elected president of his graduating class and received his medical degree in 1961. Later, Dudrick was named Intern of the Year at the Hospital of the University of Pennsylvania, where he completed his surgical residency under Jonathan E. Rhoads, M.D., and where he began pursuing the then-remote possibility of TPN (Sanchez and Daly, 2010). Further insights into the origins of TPN can be gleaned from those recent testimonials,

> Dr. Dudrick was not easily daunted nor was he afraid of hard work! Regardless of how dire the prospects of intravenously feeding patients appeared to be, Stanley Dudrick would not quit. Consider the challenge of going to the laboratory to figure this out in the 1960s. There were no substrates, there were no suitable catheters, there were no protocols, and there were no infusion pumps! This seems to have been a time when Dr. Dudrick's analytical and determined upbringing came into play. His time-intensive, meticulous experiments actually began with a pair of $2 puppies from the pound. He later expanded the effort to include a dozen beagle puppies outfitted with supplies he purchased at a local auto parts store – Pep Boys (for those who remember, Manny, Moe, & Jack). Finally, Dudrick's pioneering work included his dramatic efforts to save the life of that first human baby girl, about whom he still spoke with clear affection 50 years later!

Over a half-century career in medicine, Dudrick also taught tens of thousands of medical students and chaired the surgery departments at the University of Texas Medical School at Houston, the University of Pennsylvania, and two Yale-affiliated hospitals. He was the founder and first president of the American Society for Parenteral and Enteral Nutrition (ASPEN), a clinical research and education organization that has grown to 6,500 since Dr. Dudrick started the group in 1975.

"Today, medical science takes for granted the technique of parenteral nutrition as part of the routine armamentarium in the treatment of a wide variety of difficult medical conditions in both adult and pediatric patients. Many institutions have dedicated nutritional support teams, and an entire industry has arisen to support its use" (Sanchez and Daly, 2010). None of this might ever have happened if it weren't for the research of an inquisitive and steadfast surgical resident, who through tedious trial-and-error surmounted the myriad obstacles standing in the way of TPN.

We should all give great thanks to the work of this self-described "stubborn Pole" for his vital contribution to medical science. That having been said, Dudrick never foresaw a clear end to his tireless research. As he confessed in an early interview in *The New York Times* (Brody, 1977), when he began feeding his beagles by vein, he had "no concept of what this was going to lead to" and there was no way in his "wildest imagination would he have thought of all of the reasons for which TPN is now being used." Yet, by 1977, "he did visualize the day when TPN will be routine" as it is now. Why? "Because to be consistently good, medicine still has to be practiced with the basics, and nutrition is basic to us all." With or without pleasure, we must all eat to live – even by intravenous feeding.

REFERENCES

Brody, J. E. (1977, November 25). Hospital Deaths Being Cut by Use of Intravenous Feeding Technique. *The New York Times.* www.nytimes.com/1977/11/25/archives/hospital-deaths-being-cut-by-use-of-intravenous-feeding-technique.html

Dudrick, S. J. (2003). Early Developments and Clinical Applications of Total Parenteral Nutrition. *Journal of Parenteral and Enteral Nutrition, 27,* 291–299.

Food and Agriculture Organization of the United Nations (2019). *The State of Food Security and Nutrition in the World.* www.fao.org/3/ca5162en/ca5162en.pdf

Kleiman, D. (1983, July 13). A Young Child Lives, a Hostage to Medicine. *The New York Times.* https://timesmachine.nytimes.com/timesmachine/1983/07/13/issue.html

Misericordia Faculty Research Brochure (2013, February 20). https://issuu.com/misericordiau/docs/facultyresearchbrochure_2013

Sanchez, J. A. and Daly, J. M. (2010). Stanley J. Dudrick, MD: A Paradigm Shift. *Archives of Surgery, 145,* 512–514.

Smith, H. (2020, March 4). Stanley Dudrick, Whose Surgical Technique Feeds Those Who Can't Eat, Dies at 84. *Washington Post.* www.washingtonpost.com/local/obituaries/stanley-dudrick-whose-surgical-technique-feeds-those-who-cant-eat-dies-at-84/2020/03/04/4b47f012-5e2a-11ea-9055-5fa12981bbbf_story.html

Tappenden, K. A., et al. (2020). Tributes to Our First President and Founding Father, Stanley J. Dudrick, MD, FACS, FASPEN. *Journal of Parenteral and Enteral Nutrition, 44*, 159–171.

FURTHER MATERIAL

American Society for Parenteral and Enteral Nutrition – ASPEN (2020, January 19). *ASPEN Mourns the Loss of Its First President, Stanley J. Dudrick, MD, FACS.* www.nutritioncare.org/News/General_News/ASPEN_Mourns_the_Loss_of_its_First_President,_Stanley_J__Dudrick,_MD,_FACS/

Roberts, S. (2020, February 27). Dr. Stanley Dudrick, Who Saved Post-Surgical Patients, Dies at 84. *The New York Times.* www.nytimes.com/2020/02/27/science/dr-stanley-dudrick-dead.html

10 What's in a (Drug) Name?

> What's in a name? That which we call a rose
> By any other name would smell as sweet.

These renowned lines from William Shakespeare's *Romeo and Juliet* underscore the often-arbitrary relationship that holds between sign and significate. Of course, the sound of a word sometimes bears a striking likeness to its referent, as in the case of *onomatopoeia*: buzz, cuckoo, hiss, splash, gurgle, thump, and wheeze are familiar examples. There's also the less common case where a clear pattern prevails in the naming process. A few blocks from my boyhood home, the Burke family hung one engraved wooden shingle after another for each new blessed arrival: Bonny, Conny, Donny, Johnny, and Lonny. I'm unsure whether the Burkes ever got as far as Ronny.

One especially intriguing domain in which novel names have to be assigned is pharmacy. Take the most common of drugs, Aspirin. In 1899, Bayer derived the brand name from acetylsalicylic acid. The letter "A" stands for acetyl; "spir" stems from the plant known as *Spiraea ulmaria* (meadowsweet), which yields salicin: and, "in" was a common suffix used for drugs when acetylsalicylic acid was first synthesized.

We've come a very long way since Aspirin was named. Now, novel pharmaceuticals are marketed at an ever-increasing rate; as new drugs are approved for dispensing to the public, fresh brand names must be contrived for each. This task is both challenging and intricate. In point of fact, each drug has *three* names attached to it (Smith Marsh, 2018).

The first is the drug's *chemical* name. This name is provided by the pharmaceutical company according to detailed rules established by the International Union of Pure and Applied Chemistry. The second is the drug's *generic* name. This name is provided by the

United States Adopted Name Council in accord with the active ingredient of the drug. This designation must also be assessed and approved by the World Health Organization's International Nonproprietary Name Program. The third is the drug's *brand* name. This name is given by the pharmaceutical company and must be approved by the United States Food and Drug Administration (FDA).

Of course, most of us aren't pharmacists, so the chemical name is altogether useless. Even generic names are frequently long, complicated, and difficult to pronounce and remember. So, brand names are what primarily concern consumers. These brand names are therefore the present focus of interest. They are also central to the marketing of modern pharmaceuticals, where tens of billions of dollars in profits lie in the balance. Let's now set the stage for appreciating the byzantine business of branding prescription drugs.

As of June, 2019, there were over 19,000 prescription drug products approved by the FDA for marketing in the United States. The FDA's Center for Drug Evaluation and Research (CDER) approved fifty-nine novel drugs in 2018; this total breaks the previous record of fifty-three novel drug approvals in 1996. Among the often exotic and tongue-twisting new monikers were Biktarvy (for HIV), Erleada (for prostate cancer), Ilumya (for plaque psoriasis), Lucemyra (for opioid withdrawal), Aimovig (for migraine), Olumiant (for rheumatoid arthritis), Seysara (for severe acne), and Aemcolo (for diarrhea).

Why do these new drug arrivals have such bizarre-sounding names? To some extent, it's simply a numbers game (Collier, 2014). With so many already-branded drugs and with so many more being approved each year, it is incumbent on pharmaceutical companies to generate new names on a continuing basis. Given that there are only twenty-six letters in the English alphabet to create new brand names and 171,476 real words currently in use that must be avoided, it is small wonder that new drug names may often seem so outlandish.

Furthermore, novel brand names must be chosen with considerable care. It is critical that new names do not closely resemble previously approved names; such look-alike, sound-alike (LASA) confusions are to be scrupulously avoided (Cheng, 2018). Several salient examples of LASA mix-ups have been identified, including Actonel and Actos, Celebrex and Celexa, Keflex and Keppra, Lunesta and Neulasta, Nexium and Nexavar, Paxil and Plavix, Prozac and Prilosec, Xanax and Zantac, and Zyprexa and Zyrtec.

How problematic are LASA misunderstandings? Consider some statistics. With nearly half of all Americans taking a prescription drug each month, the Institute of Medicine has estimated that more than 1.5 million people are sickened, injured, or killed each year by errors in prescribing, dispensing, and taking medications (Scutti, 2016). With specific regard to LASA errors, at least 1 out of every 1,000 prescriptions is linked with a wrong-drug error; name confusion is suspected to be a prime culprit. Although many accidental errors do not cause notable harm, some do and demand hospital care. A few such blunders have caused death. Schultz (2013) reported that between 2009 and 2013, some 174 incidents of LASA confusions have been reported to the FDA: 9 caused critical illnesses and 16 were fatal.

All of these considerations have led a number of pharmaceutical companies to outsource the branding of drugs to so-called name engineering agencies. These firms fulfill highly specialized tasks; they both generate and select names that they hope will stimulate drug sales and also pass all of the many regulatory hurdles that must be cleared en route to final approval. Not only must our own FDA approve the brand name but so too must the European Medicines Agency and HealthCanada, because drugs represent a big international business. Experts in the industry admit that we have now entered the "regulated age of drug naming."

NAMING PROCESS: GENERATION AND SELECTION

So, just how might the naming process proceed? One especially interesting approach (reviewed by Scutti, 2016) has been offered by Denis Ezingeard, CEO of ixxéo healthcare – a pharmaceutical naming consultancy located in Geneva, Switzerland. According to Ezingeard, his "Name Engineering" model applies science to the process of branding, in order to craft drug names with the greatest potential to flourish in the global pharmaceutical marketplace. The specific source of inspiration for this model turns out to be *biological evolution*.

ixxéo's (2010) elaborate promotional materials introduce its overall approach by drawing a strong analogy between Darwinian evolution and their firm's own drug naming model. Here is their marketing pitch: "There are ingenious mechanisms that enable colonies of bird species to evolve and endure – much like the Darwin's finches that progressively adapted to ever more stringent conditions in the Galapagos Islands. Similar mechanisms drive Name Engineering, our innovative and adaptive approach to pharmaceutical naming." More specifically: "The evolutionary principles of genetic variation, adaptive radiation, natural selection, and speciation have inspired our divergent direction. These principles provide a natural framework and powerful drivers for our naming model and processes."

Of course, naming drugs is a very long way from evolving species. That's the challenge accepted by this particularly innovative consultancy. For any new drug, the ixxéo team generates several thousand prospective names. The name *generation* process is claimed to follow practices akin to jazz improvisation and modern art: "We mix original morphological elements derived from roots and word patterns across languages, to generate vast and rich assortments of name families and variants. Then, we craft visual and melodic "naming genes" and instill these subtle hooks into names – to resonate and engage in the marketplace." Few explicit details are provided, of course, in order to prevent ixxéo's proprietary trade secrets from being copied by competitors.

Then, the weeding out of unsuitable names must commence. Much of that elimination follows from the many regulatory constraints that were noted earlier. But, the final name *selection* process is one demanding consummate human skill and aesthetic sensitivity. "Both abstract paintings and names have little or no tangible content. But when artistically crafted, they can affect us in many ways, with sensations of emotional appeal that are free of contextual reference."

Here is where this specific company is believed to excel and why it has risen to prominence in the branding market; some of the most prominent names produced by its creative staff include Viagra, Angeliq, Enviage, and Certican.

Although ixxéo healthcare has paved the way for greater originality in the drug branding business, some pharmaceutical companies are still content to let computer algorithms do the work. Without sufficient screening for salience and memorability, perhaps that's why we wind up with so many weird and unpronounceable names (although one of the most difficult to pronounce names, Xalkori, had in fact gone through a rigorous and expensive selection process; see Schultz, 2013).

Considering all of this information about contemporary drug naming, one thing is patently clear: Naming drugs is big business. Drugs are costly to create; the tab can now reach as much as a half million dollars. Drugs can take up to five years to develop. And, the FDA may not give its final approval until three months before the drug is set to go on sale to the public. Failure at this point would represent a major setback in time and money.

Variation and selection are thus hard at work in this remarkable realm of commerce. It's not just for birds in the Galapagos anymore.

POSTSCRIPT

We now see that a significant portion of the cost of prescription drugs is attributable to the naming process itself: a process that has nothing whatsoever to do with the therapeutic effectiveness of the medication. When consumers complain that drugs are too expensive, who is to blame for those exorbitant prices? The answer is that, at least in part, we are!

Those advertisements that inundate the media are meant for us. When we hear about new, promising medication, we ask our physicians for those drugs – and we do so by name. The more appealing and memorable the name, the more likely we are to request it. Higher demand means greater profits. So, in this case, the answer to the opening question of this vignette "what's in a name" is "money."

REFERENCES

Cheng, C. (2018). Look-Alike, Sound-Alike Medication Names: The Role of Indications. *First Databank*. www.fdbhealth.com/insights/articles/2018-10-04-reducing-errors-from-look-alike-sound-alike-medication-names-the-role-of-indications

Collier, R. (2014). The Art and Science of Naming Drugs. *Canadian Medical Association Journal*, 186, 103.

www.ncbi.nlm.nih.gov/pmc/articles/PMC4188646/pdf/1861053.pdf

Schultz, D. (2013, August 22). With a Name Like Xalkori. *Slate*. https://slate.com/technology/2013/08/drug-name-confusion-fda-regulations-and-pharma-create-bizarre-new-names.html

Scutti, S. (2016, November 25). 'Creation Engineering': The Art and Science of Naming Drugs. *CNN*. www.cnn.com/2016/11/25/health/art-of-drug-naming/index.html

Smith Marsh, D. E. (2018). Overview of Generic Drugs and Drug Naming. *Merck Manual, Consumer Version*. Kenilworth, NJ. www.merckmanuals.com/home/drugs/brand-name-and-generic-drugs/overview-of-generic-drugs-and-drug-naming#

The Science of Naming (2010). *ixxéo Healthcare*. www.ixxeo-healthcare.com/our-approach/the-science-of-naming

11 Self-Medication by People and Animals

Google estimates that people make 1 billion searches each day as they seek information about their health. David Feinberg, the head of Google Health, reported in 2019 that some 7 percent of Google's daily searches concerned health-related matters: "People are asking us about conditions, medication, symptoms and insurance questions. In this case we are organising the world's health information and making it accessible to everyone" (Murphy, 2019).

This intense interest in healthcare is certainly to be expected. As Shakespeare's Hamlet observed in his famous soliloquy, "the thousand natural shocks that flesh is heir to" can drive us from distraction to desperation as we fervently seek some remedy for our many ailments.

Of course, the Internet was not always at our beck and call to help us minister to the illnesses and injuries we endure. "Dr. Google" has now become the first go-to resource when our wellbeing is threatened and we decide to initiate the process of self-medication – either out of expediency, necessity, or frugality (Kopp, 2019). The information to be found on the Internet may not be authoritative or prove especially useful, as it can come from a wide variety of sources. Nevertheless, uncertainly is itself unpleasant, as revealed by Hamlet's plaintive fretting over whether or not to take action against his adversaries. Valid or not, information may at the very least help allay the uncertainty of the anxious patient.

MEDICAL HISTORY AND PREHISTORY

The recorded history of medicine extends over an extremely long timeline (Hajar, 2015). Pride of place as the earliest practitioner of medicine may go to the Egyptian polymath Imhotep (circa 2686 to

2613 BCE), who is said to have diagnosed and treated as many as 200 diseases (Barton, 2016). An ancient text, believed to have been written by Imhotep, is the first known manual of surgery and trauma; it describes four dozen cases of wounds, fractures, dislocations, and tumors. However, many authors have questioned whether Imhotep ever practiced medicine, instead suspecting that a funerary cult was responsible for his inflated historical reputation (Risse, 1986).

Even more speculation revolves around the prehistory of medicine, when our ancestors could neither read nor write, thereby forcing anthropologists to rely on human remains and artifacts to tell their own tales. Without the benefit of well-developed scientific methods and theories, many writers have conjectured that prehistoric peoples would have believed that both natural and supernatural factors participate in causing and treating various maladies (Brazier, 2018). Nonetheless, we should not consider such supernatural beliefs to have rendered prehistoric peoples altogether unable to treat what ailed them. It may well have been the case that the familiar process of trial-and-error learning participated in prehistoric medicine, especially in regard to the use of medicinal herbs and other plant substances (Applequist and Moerman, 2011). This possibility is receiving surprising support from observations of self-medication in animals.

SELF-MEDICATING ANIMALS?

"Birds do it, bees do it, even educated fleas do it." No, I'm not alluding to their falling in love (my sincere apologies to Cole Porter and to uneducated fleas); rather, I'm alluding to their engaging in self-medicating behaviors.

No less than humans, animals suffer from accident and disease. Beyond the basic bodily mechanisms that promote healing and the return to normality (dubbed *homeostasis* by the renowned American physiologist Walter B. Cannon), one might wonder whether animals also self-medicate. A growing body of evidence suggests that they do, thus prompting profound questions concerning the origin of self-medicating behaviors by both animals and humans. Indeed, the science of animal self-medication has actually been given a name; it is called *zoopharmacognosy*, derived from the Greek roots zoo ("animal"), pharma ("drug"), and gnosy ("knowing") (Shurkin, 2014).

To critically assess any alleged instances of zoopharmacognosy, it would be extremely useful to have some basic ground rules to follow. Michael Huffman of the Primate Research Institute of Kyoto University is one of the world's leading authorities on the subject; in numerous publications written over three decades, Huffman has suggested several criteria for confirming the involvement of self-medication in animal behavior, especially when plant material is ingested.

According to Huffman (2016), four criteria ought to be met in order to properly establish self-medication. One should: (1) specify the illness or symptom(s) being treated; (2) distinguish the use of a therapeutic food from the organism's dietary staples; (3) document a clear positive change in health following the occurrence of self-medicative behavior; and (4) provide pharmacological evidence pinpointing the effective chemical compound(s) contained in these therapeutic agents. In practice, it is extraordinarily difficult to satisfy all of these criteria, thereby leaving considerable latitude for controversies to arise. Even more challenging is specifying how humans and animals ever came to engage in these self-medicating behaviors in the first place, encouraging a great deal of speculative theorizing.

Although the field of zoopharmacognosy is quite young, it is nevertheless very active. A full review would be out of place for our present purposes. So, let's consider some illustrative examples, with due regard to the evidence often being more suggestive than definitive.

Perhaps the most common instance of self-medication in animals is eating grass. Both dogs and cats occasionally eat grass. Most pet owners might wonder why these natural carnivores would ever do so. Many possible reasons have been proposed (Shultz, 2019).

First, grass contains folate – a B vitamin. This vitamin is essential for promoting growth and for increasing oxygen levels in the blood. Folate is also present in red meat. If a carnivore's diet happens to be deficient in folate, then eating grass can provide a necessary vitamin boost. Second, eating grass may have a laxative benefit; eating grass is especially likely if your pet has long hair or fur, which can clog its digestive tract and cause indigestion. Third, eating grass may have an emetic effect; vomiting is yet another way in which your pet can purge its stomach of foreign matter such as bones and parasites. Fourth, some have hypothesized that eating grass can relieve your pet's sore throat. Finally, although a rather indirect and perhaps implausible reason for grass eating, because dogs and cats prey on other animals that do eat grass and other greens, over evolutionary time, they too may have acquired a taste for greenery.

Beyond eating grass, many different examples of self-medication have been reported in a wide variety of species. The prevalence of self-medication is to be expected because the ability to ward off life-threatening diseases should afford a vital adaptive advantage to any species.

Consider ants (Bos et al., 2015). These animals are prone to infection by a pathogenic fungus. When infected with the fungus *Beauveria bassiana*, ants will ingest a harmful substance that uninfected ants would otherwise avoid. Why? Because such ingestion actually leads to enhanced survival of the infected ants. The authors of this interesting study conclude that therapeutic self-medication does occur in social insects. At present, the origin and mechanisms of this behavior are unknown, although the *evolutionary history of the species* is highly likely to hold key clues.

Somewhat more is known about the origins of self-medicating behaviors in ruminants such as sheep, goats, and cattle. After extensive investigation (reviewed by Villalba et al., 2014), Juan Villalba at Utah State University has come to believe that these animals engage in a trial-and-error learning process akin to elementary scientific experimentation. As was true for ants, ruminants come to eat foods rich in substances that relieve them of alimentary ailments. However, unlike the earlier example of ants, the process here involves the *prior experience of individual animals*. "There's a learning process that needs to take place in order for them to develop a preference for foods that contain those medicines" (Villalba quoted in Velasquez-Manoff, 2017).

This interesting story starts with the fact that sheep and other ruminants are frequently infected by gastrointestinal endoparasites such as nematodes: tiny worms that cling to vegetation and that negatively affect the infected animal's health, daily activities, nutritional state, and overall fitness. In response to that challenge, sheep have been seen to eat foods that contain compounds capable of treating such parasitic infections. What was needed beyond these intriguing initial observations was clear experimental evidence to properly document this phenomenon and to provide an objective account of its emergence.

Villalba et al. (2010) collected just such controlled experimental evidence. Lambs usually avoid plants containing tannins because they taste bitter. However, tannins have antiparasitic properties: They kill parasites and relieve infections. So, the researchers infected several lambs with nematode larvae and assigned them to one of two groups: the first received tannin-enriched alfalfa and the second received their customary diet of unadulterated alfalfa. The parasite counts in the stool samples of lambs that had eaten tannin-enriched alfalfa dropped compared to those that had not; this disparity therefore confirmed the medicinal value of the added tannin. When the researchers later infected the same lambs with nematode larvae a second time and gave them a choice between tannin-enriched or regular alfalfa, only the lambs that had eaten the tannin-rich alfalfa and had benefited from its antiparasitic effects chose it; the lambs that had not eaten the tannin-rich alfalfa and had not benefited from its antiparasitic effects continued to eat their regular food and continued to be infected with parasites. This study thus supported the role of individual learning in driving the lambs to select the medicinally appropriate food – a clear case of self-medication.

How lambs would ordinarily learn to do so on their own definitely merits further investigation. One promising possibility is that gastric illness may encourage lambs to vary their diet. Such variation would increase their exposure to potentially therapeutic foods – which would not otherwise be chosen because of their being distasteful.

Huffman's own primate research began in Africa. There, native peoples told him tales of wild chimpanzees engaging in a variety of self-medicating behaviors. For example, when suffering from parasitic infections, chimpanzees will eat the leaves of *Vernonia amygdalina*, a small shrub from the daisy family. This plant is relatively scarce and is unevenly distributed throughout the chimpanzees' foraging territory. The plant does not seem to be avidly eaten as food, but is occasionally consumed as medicine,

> When ingesting the pith from younger fleshy shoots of *V. amygdalina*, chimpanzees carefully remove the bark and leaves, then chew on the inner pith, extracting only the extremely bitter juice and only small amounts of fiber. The amount of pith taken at one time is insignificant from a nutritional perspective. Altogether, depending on the number of leaves ingested, this act takes anywhere from less than 1 to 8 minutes, and bitter pith chewing appears to be a one-dose treatment. Individuals have not been seen to chew bitter pith again within the same day or even within the same week. This might possibly be due to the plant's toxicity. Adult chimpanzees in proximity to a sick individual chewing *Vernonia* bitter pith very rarely show interest in ingesting the pith themselves. On the other hand, infants of ill mothers are known to taste small amounts of the pith discarded by their mothers. Despite a year round availability of the plant, use of this pith by chimpanzees is highly seasonal and rare, occurring most frequently during the rainy season peak in nodule worm infections – as expected for a therapeutic treatment. (Huffman, 2016, 6)

Huffman observed that only patently ill chimpanzees consumed *Vernonia amygdalina*; these sickened apes exhibited a sharp loss of appetite, listlessness, diarrhea, or constipation. However, after consuming the plant, they usually completely recovered from their symptoms within only 20–24 hours.

These observations are truly amazing. So too is the fact that native African tribespeople have reported to Huffman that they copied the behaviors of their chimpanzee neighbors to treat their own illnesses! Numerous ethnic groups in Africa concoct potions made from *Vernonia amygdalina* leaves or bark to treat parasitic infections, malaria, dysentery, and diarrhea. We humans may not be the insightful innovators we arrogantly believe ourselves to be; we may sometimes be mere copycats.

One more fascinating possibility is also worth noting. Recent archeological and biochemical evidence suggests that an extinct prehuman relative in our evolutionary ancestry, *Homo neanderthalensis*, may also have used the same medicinal plants to treat their maladies as we do now. According to Hardy, Buckley, and Huffman (2016), "Neanderthals survived for around 300,000 years, and were able to adapt to a wide range of environments, making them a highly successful species; we suggest that this would have been impossible had they not known what to eat in order to remain healthy and reproduce successfully, while avoiding poisoning themselves in the process" (1377).

Self-medication may thus have a long evolutionary history, which embraces human evolution, as well as the evolution of most living animal species. Clearly, this highly adaptive brand of self-medication departs dramatically from the dangerous form of escapism that is commonly said to underlie the kind of self-medication involved in drug addiction and abuse. Yet, the alleviation of any aversive symptom – whether physical or psychological – may reinforce any means that proves to be effective.

The irony that a common behavioral mechanism may underlie both self-medication and drug abuse should not be ignored. Nor can the emerging evidence that animals may not only self-medicate to *escape* from *prevailing* noxious conditions, but also to *avoid* their *future* occurrence altogether. Prophylaxis too may be in animals' weaponry against illness and infection (de Roode, Lefèvre, and Hunter, 2013).

We clearly have many more questions to answer in this extremely exciting and demanding domain. Answers to those questions may depend on our protecting tropical rainforests, where many of tomorrow's most effective medicines may come. The medicinal value of plants remains a vast, untapped resource for fighting life-threatening diseases (Suza, 2019).

REFERENCES

Applequist, W. L. and Moerman, D. E. (2011). Yarrow (*Achillea millefolium* L.): A Neglected Panacea? A Review of Ethnobotany, Bioactivity, and Biomedical Research. *Economic Botany*, 65, 209–225.

Barton, M. (2016, May 28). Imhotep – The First Physician. *Past Medical History*. www.pastmedicalhistory.co.uk/imhotep-the-first-physician/

Bos, N., Sundström, L., Fuchs, S., and Freitak, D. (2015). Ants medicate to fight disease. *Evolution*, 69, 2979–2984.

Brazier, Y. (2018, November 1). What Was Medicine Like in Prehistoric Times? *Medical News Today*. www.medicalnewstoday.com/articles/323556.php

de Roode, J. C., Lefèvre, T., and Hunter, M. D. (2013). Self-medication in Animals. *Science*, 340, 150–151.

Hajar, R. (2015). History of Medicine Timeline. *Heart Views*, 16, 43–45.

Hardy, K., Buckley, S., and Huffman, M. (2016). Doctors, Chefs or Hominin Animals? Non-edible Plants and Neanderthals. *Antiquity*, 90, 1373–1379.

Huffman, M. A. (2016). Primate Self-medication, Passive Prevention and Active Treatment – A Brief Review. *International Journal of Multidisciplinary Studies*, 3, 1–10.

Kopp, D. (2019, October 15). Cyberchondriacs? Actually, Googling Their Symptoms Makes Patients More Informed. *Newsweek*. www.newsweek.com/doctor-google-webmd-cyberchondriacs-self-diagnosing-trust-doctors-1465443

Murphy, M. (2019, March 10). Dr Google Will See You Now: Search Giant Wants to Cash in on Your Medical Queries. *Daily Telegraph*. www.telegraph.co.uk/technology/2019/03/10/google-sifting-one-billion-health-questions-day/

Risse, G. B. (1986). Imhotep and Medicine: A Reevaluation. *Western Journal of Medicine*, 144, 622–624.

Shultz, D. (2019, August 8). Mystery Solved? Why Cats Eat Grass. *Science*. www.sciencemag.org/news/2019/08/mystery-solved-why-cats-eat-grass

Shurkin, J. (2014). Animals That Self-Medicate. *Proceedings of the National Academy of Science*, 111, 17339–17341.

Suza, W. (2019, November 14). Dwindling Tropical Rainforests Mean Lost Medicines Yet to Be Discovered in Their Plants. *The Conversation.* https://phys.org/news/2019-11-dwindling-tropical-rainforests-lost-medicines.html

Velasquez-Manoff, M. (2017, May 18). The Self-medicating Animal. *The New York Times Magazine.* www.nytimes.com/2017/05/18/magazine/the-self-medicating-animal.html

Villalba, J. J., Miller, J., Ungar, E. D., Landau, S. Y., and Glendinning, J. (2014). Ruminant Self-medication against Gastrointestinal Nematodes: Evidence, Mechanism, and Origins. *Parasite, 21,* 1–10.

Villalba, J. J., Provenza, F. D., Hall, J. O., and Lisonbee, L. D. (2010). *Journal of Animal Science, 88,* 2189–2198.

FURTHER MATERIAL

Huffman, M. (2012). *Animal Self-medication* [Video]. TEDx Conference. www.youtube.com/watch?v=WNn7b5VHowM

12 Personalized Medicine
The End of Trial-and-Error Treatment?

All people are created equal. At least, so proclaims our revered *Declaration of Independence*. However, when it comes to treating illnesses, this assertion may be a serious misnomer: one size may *not* fit all. The latter proposition is the prime impetus behind a rapidly growing movement in medical care called "personalized" or "precision" medicine.

Blithely assuming that all patients are equal may lead many to suffer a protracted, exasperating, and costly "trial and error" therapeutic process, in which one treatment after another is attempted in an effort to remedy patients' many maladies. This process would usually begin by physicians' giving the most widely prescribed treatment to their patients. But, if that treatment were to prove unsuccessful, then others would be tried – in succession – until success is (if ever) achieved. Instead of this so-called hit-or-miss approach, what may be preferable is a "Mr. Rogers" tactic, in which each person's uniqueness is fully appreciated in returning the patient to health.

THE HUMAN GENOME PROJECT

Launched in 1988, the Human Genome Project provided the scientific underpinning for precisely specifying that individuality. The Shattuck Lecture delivered in 1999 by National Institutes of Health Director Francis Collins not only described the progress that had to that point been made in sequencing the human genome – the genetic material that makes us human beings and that distinguishes one person from another – but it outlined a bold vision for the future of medicine. According to that vision, genomic testing might not only help us understand and remedy the ravages of rare genetic illnesses but it might effectively fuel the even more expansive and effective field of personalized medicine to treat more common afflictions. According to Collins (1999), "Identifying human genetic variations will eventually allow clinicians to subclassify diseases and adapt therapies to the individual patient" (341).

Of course, contemporary therapies routinely prescribe drugs. Here, too, Collins foresaw that genetics might provide invaluable therapeutic information,

> There may be large differences in the effectiveness of medicines from one person to the next. Toxic reactions can also occur and in many instances are likely to be a consequence of genetically encoded host factors. That basic observation has spawned the burgeoning new field of pharmacogenomics, which attempts to use information about genetic variation to predict responses to drug therapies. (341)

Discovering new drugs as well as applying them therapeutically might also be accelerated by better understanding the genetic makeup of diverse individuals, as Collins envisaged,

> Not only will genetic tests predict responsiveness to drugs on the market today, but also genetic approaches to disease prevention and treatment will include an expanding array of gene products for use in developing tomorrow's drug therapies. These include drugs for the treatment of cancer, heart attack, stroke, and diabetes, as well as many vaccines. (341)

THE PROMISE OF PERSONALIZED MEDICINE

Since Collins' auspicious announcement, personalized medicine has been loudly and widely acclaimed. Numerous articles have been written extolling its limitless potential. An open access journal (*Journal of Personalized Medicine*) was launched in 2011, which aimed "to integrate expertise from the molecular and translational sciences, therapeutics and diagnostics," as well as to encourage "discussions of regulatory, social, ethical and policy aspects." In addition, a prominent nonprofit educational and advocacy organization was incorporated in 2004. The Personalized Medicine Coalition now actively promotes "the understanding and adoption of personalized medicine to benefit patients and the health system."

The Personalized Medicine Coalition is an especially eclectic and well-funded group comprising scientists, innovators, physicians, patients, health care providers, and insurance companies. Beyond the delivery of high-quality care to patients by physicians, the health care industry is especially well represented in the coalition. Its goals prominently include promoting the understanding and adoption of personalized medicine concepts, services, and products as well as shaping policies to support increasing financial investment in personalized medicine by addressing regulatory and reimbursement issues.

Topping off this effusive enthusiasm, on December 18, 2015, President Barack Obama authorized the $215 million Precision Medicine Initiative involving an ambitious mission, "To enable a new era of medicine through research, technology, and policies that empower patients, researchers, and providers to work together toward development of individualized care." In his remarks, President Obama promoted the aim of the initiative as "delivering the right treatments, at the right time, every time to the right person."

All of these various constituencies, resources, and authorities therefore seem to be effectively positioned for personalized medicine to usher in a fresh paradigm for delivering unsurpassed health care to patients – one based on the latest developments in human genetics and capable of radically transforming therapeutic medicine. So, after a full three decades of work, where do things now stand?

PROMISES, PROMISES

Despite the unreserved publicity energizing the movement, Joyner and Paneth (2019) have quite soberly assessed the accumulated accomplishments of personalized medicine. They found "no impact of the human genome project on the population's life expectancy or any other public health measure, notwithstanding the vast resources that have been directed at genomics" (947). Even worse, despite the absence of compelling evidence to justify the persistent lavishing of support for the program, the personalized medicine juggernaut rolls inexorably along, "Exaggerated expectations of how large an impact on disease would be found for genes have been paralleled by unrealistic timelines for success, yet the promotion of precision medicine continues unabated" (947).

Joyner and Paneth (2019) put the costs and benefits of this highly influential movement into the following stark perspective,

> In light of the limitations of the precision medicine narrative, it is urgent that the biomedical research community reconsider its ongoing obsession with the human genome and reassess its research priorities including funding to more closely align with the health needs of our nation. We do not lack for pressing public health problems. We must counter the toll of obesity, inactivity, and diabetes; we need to address the mental health problems that lead to distress and violence; we cannot stand by while a terrible opiate epidemic ravages our country; we have to prepare conscientiously for the next influenza pandemic; we have a responsibility to prevent the ongoing contamination of our air, food, and water. Topics such as these have taken a back seat to the investment of the National Institutes of Health and of many research universities in a human genome-driven research agenda that has done little to solve these problems, but has offered us promises and more promises. (948)

A REVEALING "SUCCESS" STORY

Is the outlook for personalized medicine really as bleak as these critics claim? Will success never be achieved? And, if success is achieved, then on what does it truly depend and how much might it actually cost?

As an illuminating example, consider this poignant case, reported by Gina Kolata in *The New York Times* and by Francis

Collins in the *NIH Director's Blog* in October of 2019. Mila Makovec is now nine years old and lives in Longmont, Colorado with her mother and younger brother. Since she was three years old, Mila has suffered the devastatingly debilitating symptoms arising from a form of Batten's disease: the common name given to a broad class of rare rapidly progressive neurodegenerative diseases that leave children seizure-prone, blind, cognitively and motorically impaired, and bedridden until they die prior to or soon after reaching their teenage years.

Mila's case appears to be unique because the gene that is usually necessary to produce the disorder must be inherited from *both* parents; yet, only *one* of her parents actually had that gene. Mila's particular problem resulted from an extra bit of DNA that impaired the production of a vital brain protein.

After corresponding with Mila's mother, Dr. Timothy Yu and his colleagues at Boston Children's Hospital (Kim et al., 2019) went to work to help Mila. The team first diagnosed the genetic source of Mila's disorder. Next, they began an extensive series of testing trials, some involving rodent models, to assess the effectiveness of several experimental drugs that were similar to those that had previously been developed to treat another hereditary neurodegenerative disease caused by a different faulty protein. Over the course of the better part of a year, the results of these many testing trials proved sufficiently promising for the Food and Drug Administration to approve giving the best candidate to Mila. That drug was named *Milasen* after Mila. She was given her first dose on January 31, 2018.

Since then, Mila's condition has shown measurable signs of improvement. Nevertheless, Mila still has an extremely long way to go if she is ever to become the vivacious child she had been (videos posted on Mila's Miracle Foundation website vividly portray the ravages of her disease). Soon after treatment with *Milasen*,

> Mila was having fewer seizures, and they were not lasting as long. With continued treatments, the number of seizures has diminished

so much that the girl has between none and six a day, and they last less than a minute. Mila rarely needs the feeding tube now, and is able once again to eat puréed foods. She cannot stand unassisted, but when she is held upright, her neck and back are straight, no longer slumped. Still, Mila has lost the last few words of her vocabulary and remains severely disabled. (Kolata, 2019)

For any parent whose child has been horribly devastated by disease, any sign of hope is of inestimable comfort and encouragement. Even the public at large has wished Mila, her family, and her research team every success and, through crowdsourcing, has contributed financially to her very costly care and treatment. Yet, medical progress has many yardsticks – both ethical and financial. We next turn to that process of assessment.

HOW CAN WE EVALUATE THE SUCCESS OF PERSONALIZED MEDICINE?

The first scientific paper reporting the discovery, development, and administration of a drug therapy specifically crafted for a *single* patient (Mila) was reported in the *New England Journal of Medicine* on October 24, 2019 (Kim et al., 2019). Beyond the scientific significance of that report, the first thing that may strike the reader's eye is the number of contributing authors: 48! That is a truly stunning number of individuals to have participated in the research. So too is the price tag of the project: conservatively estimated to exceed $3 million! Only the public's generous private donations could have allowed that level of care.

Published along with the Kim et al. (2019) study was a thoughtful editorial by Woodcock and Marks (2019) followed soon after by an editorial published in (*Nature Medicine* (The Cost of Getting Personal), 2019). These two editorials highlight the many unanswered ethical and financial questions that arise from such *"N-of-one"* studies. Here are just a few,

What kind of evidence justifies giving a person an experimental drug?
What safety measures are put into place to react promptly to any adverse side effects?
How does one decide what dosage and regimen are appropriate?
What measures of improvement should be used to evaluate the drug's effectiveness?
What criteria will be used for deciding when to cease seemingly ineffective treatment?
Can society afford to treat such rare conditions with ultra-personalized therapies?
Will these personalized therapies be available only to the wealthy?

Added to the earlier public health concerns raised by Joyner and Paneth, these questions make it clear that personalized medicine is still in its infancy. This highly touted end of "trial-and-error" medicine is itself tightly ensconced in its own "trial-and-error" stage of development, in which "what is true today may be obsolete in a week" (Kanter and Desrosiers, 2019, 180).

Nevertheless, these issues may not be cause for despair. Even if only a single patient can benefit from such personalized therapy, it may be possible to learn from that success and apply the basic methodology to treat other illnesses. In the future, such personalization may not only apply to the rarest of illnesses but to others where different medications may preferentially work in treating different patients. The ability to build on past successes may represent the best means by which to assess the utility of personalized medicine.

A HISTORICAL NOTE ON THE ORIGIN OF PERSONALIZED MEDICINE

With its strong connection to the Human Genome Project, the personalized medicine movement certainly enjoys the patina of modernity. It is therefore of considerable interest that a recent analysis of all of the Hippocratic texts, called the *Corpus Hippocraticum*, using original translations, suggests that the foundations of personalized medicine may actually have developed from medical precepts and practices dating from Greek antiquity (Konstantinidou et al., 2017). This noteworthy scholarly undertaking led the authors of that analysis to the following conclusions,

> The most important points located in the ancient passages were: (1) medicine [is] not "absolute," thus its directions cannot be generalized to everybody, (2) each human body/organism is different and responds differently to therapy; therefore, the same treatment cannot be suitable for everybody and (3) the physician should choose the appropriate treatment, depending on the patients' individual characteristics, such as different health status and life style (activities, diet, etc.). (601)

Thus, the fundamental concept of personalized or precision medicine – that the patient's individuality must be respected in fashioning effective treatment – is one that is firmly rooted in ancient medicine. Today's personalized medicine clearly represents the continued evolution of Hippocratic practice.

Some 2,400 years in the making represents an exceptionally long storyline. Therefore, given the very real obstacles that currently prevent personalized medicine from broadly penetrating ordinary medical practice, we should be cautious in embracing the unconstrained enthusiasm that one often reads about the promise of the movement.

Such caution represents a clear counterbalance to the pronouncement that was recently published in the journal *Personalized Medicine*, "It Is Hard to Overestimate the Impact of Precision Medicine on the Future of Healthcare" (Barker, 2017, 461). At least as compelling would be the contrary conclusion, "It Is All Too Easy to Overestimate the Impact of Precision Medicine on the Future of Healthcare." The jury is still out concerning that impact.

REFERENCES

Barker, R. W. (2017). Is Precision Medicine the Future of Healthcare? *Personalized Medicine, 14*, 459–461.

Collins F. S. (1999). Shattuck Lecture: Medical and Societal Consequences of the Human Genome Project. *New England Journal of Medicine, 341*, 28–37.

Collins F. S. (2019, October 23). One Little Girl's Story Highlights the Promise of Precision Medicine. *NIH Director's Blog.* https://directorsblog.nih.gov/tag/cln7/

Joyner, M. J. and Paneth, N. (2019). Promises, Promises, and Precision Medicine. *Journal of Clinical Investigation, 129*, 946–948.

Kanter, M. and Desrosiers, A. (2019). Personalized Wellness Past and Future: Will the Science and Technology Coevolve? *Nutrition Communication, 54*, 174–181.

Kim, J., et al. (2019). Patient-Customized Oligonucleotide Therapy for a Rare Genetic Disease. *New England Journal of Medicine, 381*, 1644–1652.

Kolata, G. (2019, October 10). Scientists Designed a Drug for Just One Patient. Her Name Is Mila. *The New York Times.* www.nytimes.com/2019/10/09/health/mila-makovec-drug.html

Konstantinidou, M. K., Karalangi, M., Panagopoulou, M., Fiska, A., and Chatzaki, E. (2017). Are the Origins of Precision Medicine Found in the *Corpus Hippocraticum? Molecular Diagnosis & Therapy, 22*, 601–606.

The Cost of Getting Personal: Editorial (2019). *Nature Medicine, 25*, 1797.

Woodcock, J., and Marks, P. (2019). Drug Regulation in the Era of Individualized Therapies. *New England Journal of Medicine, 381*, 1678–1680.

FURTHER MATERIAL

Mila's Miracle Foundation www.stopbatten.org/
Journal of Personalized Medicine www.mdpi.com/journal/jpm
Personalized Medicine Coalition www.personalizedmedicinecoalition.org/

PART III Hygiene

13 Florence Nightingale
Advancing Hygiene through Data Visualization

"Please be sure to wash your hands!" This helpful admonition is no longer reserved for careless preschoolers. During the 2019/2021 coronavirus pandemic, all of us were put on high alert to keep our hands clean lest we contaminate ourselves. Yet, the causal connection between germs and infectious diseases was slow to be made – it was only to be discovered in the middle of the nineteenth century!

The main figure in that story was, of course, Louis Pasteur (1822–1895). Regarded as the "father of microbiology," Pasteur's scientific discoveries provided strong empirical support for what is today called the "germ theory" of infectious disease and paved the way for the theory's practical application to clinical medicine.

However, many other individuals helped to develop this critical realm of medical science, including Ignaz Semmelweis (1818–1865) and Florence Nightingale (1820–1910). Both of these individuals are, among their other contributions, credited with pioneering hand washing as a means of preventing deadly infections.

Here, my main focus will be on Florence Nightingale and her notable innovations in data visualization. But, I'll begin by considering the work of her contemporary Ignaz Semmelweis. The intersection of their work is especially interesting and illuminating.

IGNAZ SEMMELWEIS: PASSIONATE BUT UNPERSUASIVE

History sometimes plays cruel tricks on its countless contributors. The work of some may initially be held in high esteem only later to fall into utter obscurity. Still others may at first be deemed to have accomplished very little or to have failed altogether in their efforts; yet, they may subsequently garner great praise for their achievements.

The latter was the case of Ignaz Semmelweis, who is now called the "Savior of Mothers" (Kadar, Romero, and Papp, 2018).

I first learned about Semmelweis in an outstanding graduate course I took in the History of Biology at Indiana University. Thanks to being featured by German philosopher Carl Hempel in his widely read 1966 book, *Philosophy of Natural Science*, the historical stock of Semmelweis soon soared – so too did my own interest in the history of science.

Ignaz Phillip Semmelweis was born in what is now Budapest, Hungary. He was the fifth of ten children in the affluent family of József Semmelweis (a grocer) and his wife Teréz Müller. Ignaz received his early education at Catholic schools, his college education at the University of Pest, and his later medical education at the University of Vienna, where he was awarded his doctor's degree in 1844.

From 1844 to 1848, Semmelweis was appointed to the Vienna Maternity Hospital – its cost-free patient care contributing to its then being the world's largest maternity facility. Semmelweis (1861/2008) relates that, because of its large size and highly diverse staff, the hospital had in 1840 been split into two subdivisions: the First Maternity Division (staffed by men) for physicians and their obstetrical students, and the Second Maternity Division (staffed by women) for midwives and their midwifery students. Purely as a matter of convenience, patients were assigned to the two clinics on alternate days. This entirely arbitrary assignment process happened quite *fortuitously* to have created a highly controlled natural experiment that was to afford Semmelweis a unique opportunity to explain a grim disparity in mortality between these maternity wards (Loudon, 2013): Over the years 1844, 1845, and 1846, deaths in the First Maternity Division averaged 8.8 percent but only 2.3 percent in the Second Maternity Division.

What could account for this almost four-fold difference in mortality? This question consumed Semmelweis, whose attention soon turned to childbed (puerperal) fever – an infection of the female reproductive organs following childbirth or abortion – as it was a prominent cause of those deaths.

Semmelweis initially envisioned a number of hypotheses to explain this vexing discrepancy, but he quickly rejected them either because they were implausible or because they could be refuted by existing evidence. He subsequently conducted a number of experiments to test other hypotheses, in one case changing the patient's delivery position in the First Division to match that customarily used in the Second Division; however, the disparity in deaths continued.

Then, the sudden, shocking death of an esteemed colleague *accidentally* provided Semmelweis with the key clue to solving the mystery. Professor of Forensic Medicine Jakob Kolletschka often performed autopsies with his students. While conducting one such autopsy, a student happened to prick Kolletschka's finger with the same knife that had been used in an autopsy. Kolletschka soon fell ill and died of bilateral pleurisy, pericarditis, peritonitis, and meningitis – the same panoply of symptoms that was exhibited by hundreds of deceased maternity patients. Semmelweis hypothesized that invisible cadaverous particles had entered Kolletschka's bloodstream and caused him to succumb to the same infection that led to the childbed fever that claimed the lives of women in his First Maternity Division. This would prove to be the controversial conclusion that his contemporaries would adamantly reject, because they believed there to be as many as thirty different causes of childbed fever (Kadar et al., 2018).

Semmelweis would, of course, have to establish precisely how these cadaverous particles came to infect women in the First Maternity Division but not in the Second Maternity Division. Doing so was straightforward. Due to the anatomical emphasis of the Vienna Maternity Hospital, he and other physicians as well as their medical students and assistants frequently touched cadavers; the midwives and their students never did so. So, it was paradoxically the case that the very medical personnel who were caring for the patients in the First Maternity Division were actually infecting those women with the potentially fatal disease!

Semmelweis' final step was to take effective measures to completely cleanse the hands of those contacting the hospital patients. Simple hand washing with soap seemed insufficient to remove all of the clinging cadaverous particles – as was unmistakable by the foul smell that lingered for a long while on physicians' hands. After systematically experimenting with a variety of different disinfecting solutions, Semmelweis instituted routine chlorine handwashing in May of 1847. Quoting Semmelweis (1861/2008): "Consequently, mortality in the first clinic fell below that of the second. I therefore concluded that cadaverous matter adhering to the hands of the physicians was, in reality, the cause of the increased mortality rate in the first clinic" (7).

All of these remarkable results and analyses ought to have immediately elevated Semmelweis to a place in the pantheon of medical science. However, this was not to be the case. Not only were Semmelweis' ideas of disease infection and prevention roundly rejected but the reflexive tendency to reject any new findings or theories that contradict prevailing opinion was later to acquire his name – the so-called Semmelweis Reflex.

Many authors including Cunningham (2015), Larson (1989), Loudon (2013), Tulodziecki (2013), and Zoltán (2020) have highlighted a number of personal and professional failings that, for decades, would relegate Semmelweis to the backwaters of history. Larson (1989) explained these failings in the following fashion,

> Despite his dramatic results and the fact that he carefully documented and verified his findings over the years, his discovery was not widely accepted among his medical peers. Animosity against Semmelweis was intensified by the fact that he was apparently a rather tactless, single-minded, perhaps pompous and fanatical personality. He was also a foreigner who spoke German very poorly. When invited to present his findings to the Vienna Medical Society, he initially refused, and did not even publish his results for 13 years. When he finally did summarize his work, the

written product was quite polemical and communicated a resentful, angry tone. He felt that the "truth" of his work was so unquestionable that it should speak for itself. (95)

Loudon (2013) further elaborated the reasons why Semmelweis' book was so negatively received,

> The treatise of over 500 pages contains passages of great clarity interspersed with lengthy, muddled, repetitive and bellicose passages in which he attacks his critics. No wonder that it has often been referred to as "the often-quoted but seldom-read treatise of Semmelweis". When he wrote the treatise, Semmelweis was probably in the early stages of a mental illness that led to his admission to a lunatic asylum in the summer of 1865, where he died a fortnight later. (461–462)

There is one more credible reason why the "truth" in Semmelweis' treatise may not have spoken so loudly for itself: The invaluable findings it held were tediously presented in table after table containing number after number. Important trends may have been difficult to discern and to appreciate; they simply didn't grab his readers' attention. With rising concern over public health crises, we see that today's readers are demanding more visually arresting information to help them make sense of the trends in infection and remediation. This is where the work of Florence Nightingale comes into sharp focus.

FLORENCE NIGHTINGALE: CONVINCING AS WELL AS VEHEMENT

"Dry statistical facts." Who is not put off by lengthy reports that dull the mind with innumerable numerals? We crave a different, more vibrant way to compellingly convey our ideas to one another. Wouldn't "colorful diagrams" be vastly superior?

Florence Nightingale (see her portrait, circa 1873, by Duyckinick) – the widely fêted founder of modern nursing – made precisely these points in her letter to Mr. Sidney Herbert on December

19, 1857 (McDonald, 2010, has archived Nightingale's engrossing and bountiful correspondence in multiple volumes). Nightingale's follow-up mail of December 25, 1857, further developed her view on this matter and included a statistical appendix containing several printed tables in double columns. Although the appendix held empirical findings of great value, Nightingale did not believe that people of real authority would even read it, contending that "none but scientific men ever look into the appendix of a report."

To contribute to more widespread understanding of the important findings in the appendix, Nightingale also added several painstakingly crafted "diagrams." Rather than lifeless numbers and words, she believed that such highly *visualized data* would "affect thro' the Eyes what we fail to convey to the public through their word-proof ears." Nightingale hoped that her new visual diagrams would better enable the "vulgar public" to appreciate and embrace the rich documentary evidence that she had so assiduously assembled.

Who was the "vulgar public" whose opinion Nightingale and Herbert were so ardently trying to influence? Among others, it included Queen Victoria, Prince Albert, all of the crowned heads in Europe, all of the commanding officers in the British Army, all of the regimental surgeons and medical officers at home and abroad, as well as all of the newspapers, reviews, and magazines in the United Kingdom. Talk about leaving no stone unturned!

What was the invaluable evidence that was contained in those tables and diagrams? During and even after the fateful Crimean War (1853–1856), British soldiers suffered from dangerously deplorable conditions in military hospitals at home and abroad. Soiled bedsheets, hospital gowns infested with parasites, and rodents scurrying under the soldiers' cots were all too common. Newspaper reports decried these scandalous circumstances, prompting a public outcry for immediately redressing them. Even so, if those conditions were ever to be substantially improved, then powerful persuasion at the highest level was going to be essential in order to secure the requisite financial support and official authorization for implementing the many desired changes.

Nightingale's tables and diagrams were to serve as evidentiary grounds for establishing two key facts: first, that during the Crimean War, more deaths resulted from diseases such as cholera, typhus, and dysentery caused by poor hygiene than from combat injuries; and, second, that enhanced sanitary techniques successfully and dramatically decreased deaths from those diseases in military hospitals (Gupta, 2020).

Given these compelling facts, we must now ask: As of the year 1857, who were Florence Nightingale and Sidney Herbert? And, how did the two come to join forces in order to advance the quality of medical care for British veterans? Let's first consider the lesser known Herbert before focusing on Nightingale – the central figure of our discussion.

Sidney Herbert (1810–1861) was a British politician, a social reformer, and a dear friend and confederate of Florence Nightingale. Legend has it that Nightingale worked Herbert to an untimely demise because of the intense pressure she applied in lobbying for hospital reform; the truth is that he succumbed to chronic kidney disease. More credibly and importantly, it has been contended that we would never know the name Florence Nightingale or celebrate her remarkable achievements without the collaboration of Sidney Herbert (Foster, 2019).

The son of English and Russian nobility, Herbert was educated at elite academic institutions: Harrow School in London and Oriel College of the University of Oxford. Herbert later entered the House of Commons in Parliament in 1832, he was appointed Secretary to the Admiralty from 1841 to 1845, and he served as Secretary at War in the cabinet of Prime Minister Robert Peel from 1845 to 1846. During the Crimean War, Herbert held that same position in the cabinet of Prime Minister George Hamilton-Gordon from 1852 to 1854.

On October 14, 1854, Herbert officially requested that Nightingale lead a contingent of female nurses to care for the many sick and wounded British soldiers who were housed in the Scutari Barracks in the Üsküdar district of Istanbul, Turkey. Knowing that the troops' medical care and facilities were appallingly deficient, Herbert was adamant that something drastic had to be done to relieve the soldiers' senseless suffering. This effort was to be the first of several later reforms to bring improved medical care, initially, to the military and, later, to the general public.

Herbert and Nightingale had first met in 1847. After Herbert left government in 1846, when Peel's cabinet resigned, he married Miss Elizabeth à Court. The couple then took an extended continental honeymoon where, in Rome, they *chanced* to meet Nightingale through mutual friends: Charles and Selina Bracebridge. Thereafter, Nightingale found her growing friendship with the Herberts richly gratifying and inspirational. She especially admired the couple's philanthropic efforts to help the plight of the underprivileged. In return, the Herberts were impressed with Nightingale's striking character and passionate commitment to nursing.

So, by 1854, Herbert was entirely certain that Nightingale was the perfect person to lead this medical mission of mercy (Cook, 1914). Herbert's letter to Nightingale proclaimed there to have been "but one person in England ... who would be capable of organising and superintending such a scheme. Your own personal qualities, your knowledge, and your power of administration, and among greater things your rank and position in society, give you advantage in such a work which no other person possesses" (Stanmore, 1906, 339–340). History proves Herbert to have been absolutely correct!

Merely one week later, Nightingale, her team of ten Roman Catholic nurses, eight Anglican sisters, six nurses from St. John's Home, and fourteen nurses from English hospitals were on their way to Scutari. Nightingale herself was officially named the "Superintendent of the female nursing establishment in the English General Military Hospitals in Turkey." And, within just six months, all signs pointed to the salutary results of the mission she commanded; most impressively, the mortality rate of hospitalized

British soldiers fell from 42.7 percent to 2.2 percent. Some of Nightingale's specific improvements included increasing patient distancing by separating hospital cots by three feet, improving ventilation, removing the Calvary horses stabled in the hospital basement, flushing the hospital sewers several times daily, disinfecting the toilets and plumbing with peat charcoal – and, of course, rigorous washing of hands.

Perhaps outshining her many programmatic achievements, Nightingale was soon publicly acclaimed for her noble and caring service as the "Lady with the Lamp," because she would personally comfort ailing soldiers in the dark of night holding a small Grecian oil lamp in her hand. The celebrity status of this "ministering angel" was not lost on Nightingale in her many later public relations efforts.

After the war, Nightingale returned to England in 1856, but now in ill health; while working in Turkey, she herself had contracted a seriously debilitating disease (most likely brucellosis), which made it difficult for her to travel far from home. Nevertheless, Nightingale and Herbert – incensed by the dreadful medical treatment of British troops – carried forward the movement to dramatically improve the Army Health services.

Nightingale was extremely confident that it had been the introduction of enhanced sanitation procedures, including handwashing (Bates, 2020), that led to the marked drop in soldiers' deaths in Scutari. But, how could she best make that case to the people who mattered? This is where Nightingale's determination to effect positive social change acquired an innovative tactical weapon – *data visualization*.

Just prior to penning the two letters to Herbert noted earlier, Nightingale had expressed intense frustration with the status quo; yet, she considered that consternation as motivating her to develop a new, more powerful process of persuasion. As she wrote to Herbert on August 19, 1857, "whenever I am infuriated, I revenge myself with a new diagram." Here, we clearly see that Nightingale could be every bit as exasperated as Semmelweis by people's resolute resistance to her ideas. However, unlike Semmelweis, Nightingale was tactful and strategic rather than disrespectful and unruly in pursuing her objectives.

A key part of Nightingale's evolving political strategy was that she would not embark on her quest alone; she would solicit the knowledge and support of respected allies (Knopf, 1916). William Farr turns out to have been yet another *fortuitous* acquaintance of great significance; he played a key part in developing Nightingale's statistical and graphical expertise. She met Farr early in 1857 at a dinner party with mutual friends Colonel and Mrs. Tulloch, who had also been posted in Scutari along with Nightingale during the war.

Farr was the chief statistician and epidemiologist in the Registrar-General's Office and is widely regarded to be one of the founders of medical statistics. In 1852, Farr had used so-called circular or polar plots to visually represent the relation between weekly temperature and mortality in London from 1840 to 1850. Nightingale found Farr's graphical representations to be particularly persuasive and closely consulted with him as she tried numerous variations of those plots and other graphical portrayals to address the death rates that she had calculated from her Crimean War data (Howe, 2016).

Nightingale was a most willing and able pupil. Her correspondence with Farr revealed that a quite cordial and collegial relationship had grown between them (Howe, 2016), with Farr admiringly signing his letters to her, "I have the honour to be your very faithful servant."

In May of 1857, Nightingale sent Farr several diagrams depicting the death rates from Scutari. He replied, "Dear Miss Nightingale. I have read with much profit your admirable observations. It is like light shining in a dark place. You must when you have completed your task – give some preliminary explanation – for the sake of the ignorant reader."

Nightingale did just that in November of 1857 in the form of a written speech, which Farr found superlative. This was his fulsome appraisal, "This speech is the best that was ever written on diagrams or on the Army. Demosthenes [the great Greek statesman and orator of ancient Athens] with the facts before him could not have written or thundered better." Although poor health prevented Nightingale from orally reciting the text, her writings and diagrams were extensively expanded into the 850-page volume *Notes on Matters Affecting the Health, Efficiency, and Hospital Administration of the British Army* (1858).

The *Notes* were printed at Nightingale's own expense for private circulation among the most influential members of the "vulgar public," including Queen Victoria! Yet, Nightingale's worst fear that Victoria's "eyes would glaze over" while perusing the book's most meaty material never came to pass. Indeed, the Queen was so enthralled with the book and especially its gorgeous graphic illustrations that she lavished effusive praise on Miss Nightingale, "Such a head! I wish we had her at the War Office." With the blessing of Queen Victoria and Prince Albert, Nightingale promptly "procured support for a Royal Commission on the health of the army," which was later to reform and revolutionize military medicine and broader sanitation procedures (Grace, 2018).

Howe (2016) has critically assessed the remarkable impact of that written volume and particularly its innovative "Nightingale rose" diagram (see a reproduction at https://upload.wikimedia.org/wikipedia/commons/1/17/Nightingale-mortality.jpg) as follows,

> Her rose diagram was so easy to understand it was widely republished. Ultimately this striking visualisation and the accompanying report convinced the government that deaths were preventable if sanitation reforms were implemented in military hospitals. Nightingale's work provided a catalyst for change, driving better and cleaner hospitals and the establishment of a new army statistics department to improve healthcare.

It is, nevertheless, of special significance for us to appreciate that not all of Nightingale's graphical attempts were as successful as her "rose." She actually experimented with many variations before she selected the best depictions. Nightingale prodigiously created bar charts, stacked bars, honeycomb density plots, and area plots (Andrews, 2019). Most revealingly, Nightingale's inventive "batwing" plots were also visually striking, but they seriously distorted the results by over representing the magnitude of large scores. Her later "rose" plots proved to be equally eye-catching and far more accurate (de Sá Pereira, 2018). It was to be her "rose" plots that decisively advanced the case for those enhanced medical facilities and practices that Nightingale and Herbert so fervently sought.

Although those "rose" plots would soon become iconic and revolutionary in data representation, it merits repeating that these historical data visualizations did not occur to Nightingale in a sudden spark of inspired insight. Quite the contrary, she very plainly "arrived at her iconic rose chart through trial and, even, error" (Gupta, 2020). Sure enough, the Law of Effect played a key part in their development.

Before delving more deeply into Nightingale's personal history to help us better understand how she became such an able ally of Herbert in promoting health care reform, we should take stock of Nightingale's utilization of diagrams as potent instruments of persuasion.

In an article celebrating the bicentennial of Nightingale's birth and her contribution to data science, Hedley (2020) places special stress on the deployment of visual representation as *rhetorical* in nature – facts arguably "speaking for themselves."

> Florence Nightingale's diagrams of mortality data from the Crimean War have become icons of graphic design innovation. But what made Nightingale's graphs particularly iconic was their powerful use of visual rhetoric to make an argument about data. This quality was fundamental to how her work was an outlier in its time and place. (26)

Reflecting on Nightingale's serious concern that numerical data alone would fail to persuade, Hedley (2020) details just what is involved in the data visualization process and the numerous factors that contribute to convincing communication. Hedley's comments further magnify the outstanding and original contribution that Nightingale made to data science,

> The introduction of a visual model is the development of an interpretation of the data. The visualiser perceives a pattern among data instances, a causal or correlative relationship between variables that is meaningful in some way. Viewed in this light, a data graphic's characteristics can be understood as rhetorical techniques selected by the visualiser to help foreground a pattern they perceive in the data. This selection and curation process is tantamount to presenting an argument about the data's meaning. The visuality of the information, including design features such as shape, colour, size of components, and spatial layout, is what makes this argument accessible and persuasive, even to audiences unfamiliar with the details of the original data set. What makes Nightingale's polar area graphs innovative is their [skillful] deployment of visual rhetoric to drive home an evidence-based argument to an audience that included both political elites and a broader public. (27–28)

FLORENCE NIGHTINGALE: PRIVILEGED UPBRINGING

Of course, it is essential that we inquire just how this young Victorian woman ever found her way to the fields of nursing and data visualization. Here, we enjoy an abundance of biographical riches to mine for valuable clues (e.g., Cook, 1914; Knopf, 1916; Magnello, 2010; Schuyler, 2020), as Florence Nightingale is a truly famous and revered public figure.

Very English Nightingale was actually born in Italy – the city of Florence to be exact. Florence was the younger daughter of moneyed English parents, William and Fanny. The Nightingales were also extremely well-connected and among the "'ton," the French term commonly used to denote British high society.

Nightingale's Unitarian family included intellectually inquiring free-thinkers who fully endorsed women's education. Both Florence and her elder sister Parthenope were properly schooled in reading and writing by their governess. Several tutors also taught the sisters arithmetic, geography, botany, and French, as well as drawing and piano. Their father – a graduate of Trinity College, Cambridge – further enriched his daughters' education by providing them with the equivalent of a university education at home, teaching them mathematics, history, philosophy, classical literature, and English composition, as well as Latin, Italian, and Greek.

From an early age, Florence exhibited an extraordinary capacity for collecting and methodically recording all sorts of data, later reflecting that, "I had the most enormous desire of acquiring." Indeed, Florence documented her extensive shell collection with precisely drawn tables and lists; she even entered information about the fruits and vegetables from her garden into tidy tables. Florence was most assuredly not one for idle diversion; she was extremely serious and studious, later confessing that: "I thought of little but cultivating my intellect."

It further proved propitious that, within her family's illustrious circles, Nightingale met many Victorian literati, including the mathematician Charles Babbage; his envisioned "analytical engine" is widely considered to be the forerunner of the modern electronic computer. At 20, Florence began receiving two-hour tutoring in statistics from a Cambridge-trained mathematician. Florence was so enamored of mathematics that, when fatigued, she claimed to find a peek at a table of numbers "perfectly reviving" (Thompson, 2016).

Also, as a child, Florence intensely wished to tend the sick. When she was 17, Florence finally announced that she wanted to become a nurse. This announcement mortified many in her family, because nursing was then considered to be a lowly job for servants – one not at all suitable for well-bred ladies. Nightingale later recalled that her daydreams focused chiefly on hospitals, believing that these vivid fantasies signified that "God had called her to Him in that way." This divine calling freed Nightingale from the conventional expectations of marriage and family, which she dismissed as involving frivolous pursuits and domestic drudgeries.

Instead, Nightingale felt compelled to follow more serious intellectual and religious aspirations. "Nightingale regarded science and statistics as a substitute religion; statistics was for her 'the most important science in the world'" (Magnello, 2010, 19). Quite of her own devising, "Nightingale proposed a form of religion in which human beings actively contributed to the realisation of God's law through their work. Statistical laws provided her with a viable pathway that could reveal God's providential plan" (Magnello, 2010, 19).

One more critical facet of Nightingale's upbringing must be underscored: it relates to Herbert's firm conviction that her "rank and position in society" perfectly suited her heading the mission to Turkey. The Nightingales' social and governmental connections afforded Florence with matchless political opportunities. Her maternal grandfather, William Smith, represented Norwich in the House of Commons for almost fifty years. Her neighbor Henry John Temple served as Prime Minister from 1855 to 1858. Meeting these men as a girl "made it possible for her to later get the support from the inner circles of Parliament and Whitehall for her statistically designed sanitary reforms in military and civilian hospitals. Nightingale understood that one individual alone could not have as much impact on making radical changes in people's life without an official imprimatur from the government" (Magnello, 2010, 18).

This then was Florence's background as she was about to embark on young womanhood. She had, of course, also acquired the full gamut of etiquettes and domestic skills expected of advantaged Victorian girls. She had even been formally presented to Queen Victoria in May of 1839 (Dossey, 2000). Yet, despite all of her interests and eccentricities, Florence continued for the next few years to live with her family at their Derbyshire and Hampshire estates. Florence also routinely went on family holidays, taking every opportunity to visit hospitals in Paris, Rome, and London, where she took elaborate notes and tabulated relevant statistics.

A highly eligible young woman – intelligent, attractive, and wealthy – Florence nevertheless fended off potential suitors, even declining a serious marriage proposal in 1849. Appreciating that her future lay in nursing, Florence's father ultimately relented and, in 1850, permitted her to train as a nurse in the Kaiserswerth district of Düsseldorf, Germany under the tutelage of brothers Friederike and Theodor Fliedner at what is now fittingly named the Florence Nightingale Hospital. She embarked on further training at the Maison de la Providence in Paris. Unfortunately, Parthenope found Florence's absence so upsetting that she suffered a nervous breakdown in 1852, prompting Florence to return home to care for her sister.

At long last, in 1853, the breakthrough came: Florence finally achieved independence from the life her family expected of her and was appointed Superintendent of the Hospital for Invalid Gentlewomen in Harley Street in London. Nightingale had now become a nurse. Of course, an even higher charge was soon to arrive from Herbert in 1854.

FLORENCE NIGHTINGALE: PASSIONATE STATISTICIAN

Known worldwide as the "Lady with the Lamp," this remarkable woman was given another lesser known nickname, the "Passionate Statistician." Sir Edward T. Cook, Florence Nightingale's first biographer, fittingly joined her passion for nursing with her passion for statistics. But, recognition for her statistical innovations did not have to wait that long; in 1858, Nightingale became the first woman elected as a Fellow of the Royal Statistical Society. The American Statistical Association named her an honorary member in 1874. And, Karl Pearson – developer of the Pearson product-moment correlation coefficient – recognized Nightingale as a "prophetess" in the development of applied statistics.

Even now, Florence Nightingale's impact on statistics and data visualization is celebrated. On July 15, 2019, the Data Visualization

Society (DVS) announced its new journal dedicated to publishing articles on current issues in data visualization for professionals and the general public. Its name – *Nightingale*!

FLORENCE NIGHTINGALE: A PREPARED MIND

Earlier, I mentioned Louis Pasteur as a medical pioneer whose research dramatically advanced the case for the germ theory of disease. Pasteur is also famous for a remark he made during his Inaugural Address as the newly appointed Professor and Dean at the opening of the new Faculté des Sciences at the Université de Lille on December 7, 1854 – less than two months after Herbert officially requested that Nightingale lead a corps of nurses to Turkey. Pasteur's quotation, "chance favors only the prepared mind."

In all likelihood, no one could have been as prepared as Florence Nightingale to head that mission. But, preparation strongly implies a foreseeable end or result. Nothing in Nightingale's upbringing could have anticipated her specific interests in collecting, statistics, and nursing. Nor can Nightingale's chance meetings with Herbert and Farr have predicted her commanding that mission to Turkey or petitioning the Queen for hospital reform.

Some might invoke Divine Providence to make sense of all of this. I would not. Just like Nightingale's rose diagrams, I believe that the facts in this case "speak for themselves." Context, consequence, and coincidence can and do combine to produce remarkable results that radically change the course of human history. They rarely yield such salutary outcomes, but in this case they most definitely did. That is precisely why we celebrate the life and work of Florence Nightingale.

A travel tip: If you should travel to London, then do make your way to Waterloo Place, at the junction of Regent Street and Pall Mall, where the Crimean War Memorial is located. There you will find statues of two heroes of the campaign: Florence Nightingale is on the left and Sidney Herbert is on the right.

REFERENCES

About Ignaz Semmelweis

Cunningham, C. (2015, March 25). Communication: Semmelweis vs Florence Nightingale. *Systemic Views on Healthcare.* https://chacunningham.wordpress.com/2015/03/25/communication-semmelweis-vs-florence-nightingale/

Hempel, C. G. (1966). *Philosophy of Natural Science.* Englewood Cliffs, NJ: Prentice-Hall.

Kadar, N., Romero, R., and Papp, Z. (2018). Ignaz Semmelweis: The "Savior of Mothers." On the 200th Anniversary of His Birth. *American Journal of Obstetrics & Gynecology, 219,* 519–522.

Larson, E. (1989). Innovations in Health Care: Antisepsis as a Case Study. *American Journal of Public Health, 79,* 92–99.

Loudon, I. (2013). Ignaz Phillip Semmelweis' Studies of Death in Childbirth. *Journal of the Royal Society of Medicine, 106,* 461–463.

Semmelweis, I. (2008). The Etiology, Concept and Prophylaxis of Childbed Fever (Excerpts). *Social Medicine, 3,* 4–12. (Original work published 1861)

Tulodziecki, D. (2013). Shattering the Myth of Semmelweis. *Philosophy of Science, 80,* 1065–1075.

Zoltán, I. (2020). Ignaz Semmelweis: German-Hungarian Physician. *Encyclopaedia Britannica.* www.britannica.com/biography/Ignaz-Semmelweis#ref268466

About Florence Nightingale

Andrews, R. J. (2019, July 15). Florence Nightingale Is a Design Hero. *Medium.* https://medium.com/nightingale/florence-nightingale-is-a-design-hero-8bf6e5f2147

Bates, R. (2020, March 23). Florence Nightingale: A Pioneer of Hand Washing and Hygiene for Health. *The Conversation.* https://theconversation.com/florence-nightingale-a-pioneer-of-hand-washing-and-hygiene-for-health-134270

Cook, E. (1914). *The Life of Florence Nightingale.* London: Macmillan.

de Sá Pereira, M. P. (2018). Representation Matters. *Torn Apart: Reflections.* http://xpmethod.columbia.edu/torn-apart/reflections/moacir_p_de_sa_pereira_2.html

Dossey, B. M. (2000). *Florence Nightingale: Mystic, Visionary, Healer.* Philadelphia: Lippincott, Wilkins, Williams.

Duyckinick, E. A. (Artist). (circa 1873). *Portrait of Florence Nightingale* [Painting]. Retrieved October 19, 2020, from https://commons.wikimedia.org/wiki/File:Florence_Nightingale.png

Foster, R. E. (2019). *Sidney Herbert: Too Short a Life.* Gloucester: Hobnob Press.

Grace, S. (2018). Nightingale-1858. *The Art of Consequences.* https://edspace.american.edu/visualwar/nightingale/

Gupta, S. (2020, May 10). Florence Nightingale Understood the Power of Visualizing Science. *Science News.* www.sciencenews.org/article/florence-nightingale-birthday-power-visualizing-science

Hedley, A. (2020, April). Florence Nightingale and Victorian Data Visualisation. *Significance Magazine.*

Howe, K. (2016, August 12). "Like Light Shining in a Dark Place": Florence Nightingale and William Farr. *British Library: Science Blog.* https://blogs.bl.uk/science/2016/08/florence-nightingale-and-william-farr.html

Knopf, E. W. (1916). Florence Nightingale as Statistician. *Publications of the American Statistical Association, 15,* 388–404.

Magnello, M. E. (2010). The Statistical Thinking and Ideas of Florence Nightingale and Victorian Politicians. *Radical Statistics, 102,* 17–32.

McDonald, L. (2010). *Florence Nightingale: The Crimean War.* Volume 14 of the Collected Works of Florence Nightingale. Waterloo, CA: Wilfrid Laurier University Press.

Nightingale, F. (1858). *Notes on Matters Affecting the Health, Efficiency, and Hospital Administration of the British Army.* London: Harrison and Sons.

Schuyler, C. (2020). Florence Nightingale. In Nightingale, F. *Notes on Nursing: Commemorative Edition.* Philadelphia, PA: Wolters Kluwer. www.google.com/books/edition/Notes_on_Nursing/8Px9DwAAQBAJ?hl=en&gbpv=1&dq=constance+schuyler+florence+nightingale&pg=PT23&printsec=frontcover

Stanmore, L. (1906). *Stanley Herbert, Lord Herbert of Lea: A Memoir.* New York: Dutton.

Thompson, C. (2016, July). The Surprising History of the Infographic. *Smithsonian Magazine.*

14 Taking Mental Floss to Dental Floss

> Mental floss: Changing your thinking to get rid of useless information
>
> Urban Dictionary

"You don't have to floss all of your teeth, just the ones that you want to keep." This whimsical saying offers accepted advice from dental professionals: Flossing your teeth after eating will most assuredly improve your dental health.

Now, consider this rather surprising news flash: Flossing your teeth may significantly lessen the odds of your developing Alzheimer's Disease! The reason for this unforeseen windfall from flossing is that it reduces the ravages of the periodontal disease gingivitis. The bacterium (*Porphyromonas gingivalis*) that causes gingivitis is actually able to move from your mouth to your brain. That bacterium produces a protein that destroys the brain's nerve cells, ultimately leading to the many devastating cognitive disfunctions that are associated with Alzheimer's (Dominy et al., 2019).

Accompanying that good medical news is the good economic news that sales of dental floss are on the rise. The global dental floss market reached $2.75 billion in 2018 and it is projected to reach $3.90 billion by 2025 – a compound annual growth rate of 5.12 percent.

If he were still alive, then Levi Spear Parmly (1790–1859) would definitely be delighted with these developments. Called the "Father of Floss" (Duenwald, 2005), Parmly is credited with having invented dental floss. Yet, this familiar product was never expressly devised to combat either gingivitis or Alzheimer's disease; rather, Parmly was primarily concerned with preventing tooth decay. This corrosive process begins, he suggested, with food particles that become wedged in the gaps between the teeth.

Writing in his most frequently cited book and only the fourth volume on dentistry to be published in America – *A Practical Guide to the Management of the Teeth: Comprising a Discovery of the Origin of Caries, or Decay of the Teeth, with Its Prevention and Cure* – Parmly (1819) observed that "the interstices and irregularities of the teeth afford a lodgement for whatever is taken into the mouth; and no contrivance HITHERTO DISCOVERED can, from these parts, remove the accumulation" (70). After extensive experimentation and considerable deliberation, Parmly put together an all-purpose tooth-cleaning kit,

> This apparatus consists of three parts, contained in a small case, with a dental mirror, fit for the toilet or the pocket.
>
> The first part to be used is the brush. It is made hollow in the middle, to embrace every part of the teeth, except the interstices; and thus, at one operation, the top (a part hitherto entirely neglected), the outer and inner surfaces are completely freed from all extraneous matter. The second part is the dentifric polisher, for removing roughness, stains, &c. from the enamel, and restoring to the teeth their natural smoothness and colour. The third part is the *waxed silken thread*, which, though simple, is the most important. It is to be passed through the interstices of the teeth, between their necks and the arches of the gums, to dislodge that irritating matter *which no brush can remove*, and *which is the real source of disease.* (71–72)

Parmly's focus on preventive dentistry (some calling him the Father of Preventive Dentistry and others calling him the Apostle of Dental Hygiene, Sanoudos and Christen, 1999) is truly admirable and remarkable, given the nascent state of dentistry in his day. "Dental practice at that time was fairly primitive, and consisted primarily of extracting teeth with a turn key, cleaning the teeth with scalers, and removing caries with files and filling cavities with tin foil. Dentures were carved from ivory or bone, but swaged gold plates were beginning to be constructed, both in Europe and in America" (Chernin and Shklar,

2003, 16). So, just who was Parmly and how did he come to break such fresh ground in the realm of oral hygiene?

Levi Spear Parmly was born in Braintree, Vermont. He was one of nine children. Of the five sons, four became dentists: Levi, Eleazar, Jahial, and Samuel. Levi's father was a farmer; however, Levi vowed never to follow in his father's agrarian footsteps, finding hoeing potatoes to be unacceptably arduous. At the age of 22, Levi is said to have angrily left his father's farm proclaiming, "this is the last row of potatoes I'll ever hoe."

Accompanied by his dog and playing his violin to help pay for his wayfaring travel, Levi trekked to Boston, where he began work as a dentist's apprentice. By 1815, Parmly had learned his craft well and, after moving north to Canada, he began practicing dentistry in Montreal and later in Quebec City. It was during this period that his written work began to tout the merits of flossing. Subsequent to Parmly's publishing his major work in 1819, the peripatetic dentist practiced in London, Paris, New York City, and mostly in New Orleans.

Unfortunately, we know virtually nothing about the process by which Parmly developed the method and materials for flossing. One possibly apocryphal anecdote suggests the passion and determination with which Parmly pursued his dental inquiries. Soon after the bloody "War of 1812" Battle of Lundy's Lane, on July 25, 1814, Levi and his brother Jahial are purported to have wandered the Niagara Falls, Ontario battlefield collecting teeth from the slain soldiers to study. Those teeth might have provided Parmly with invaluable information about the mechanisms of tooth decay. Whatever the truth of this somewhat macabre tale, it is certain that Parmly made exceptional contributions to the science and practice of dentistry.

There are, to be sure, hints that Parmly may not have been the first person to devise flossing as a means of removing food matter from the gaps between the teeth. Such removal must have been a common chore for as long as humans have existed as a species – perhaps even before then. As long ago as 3,500 BCE, the ancient Egyptians used sticks with frayed ends to clean their teeth and gums, using a powder comprising rock salt, mint, and pepper; these sticks might have made some headway in removing food lodged between the teeth. Assyrian cuneiform medical texts from 3,000 BCE mention various methods for teeth cleaning; they used porcupine quills and bird feathers for cleaning teeth and the gaps between them. Toothpicks from this same time period have been discovered from multiple archeological sites in Mesopotamia. In addition, fossilized teeth from humans' hominid ancestors bear the scars of sharp objects, which may have been used for cleaning purposes.

But, the most thought-provoking leads concerning the origins of flossing come from living nonhuman primates – namely, macaque monkeys! Three different macaque species – in Thailand (*Macaca fascicularis*), Japan (*Macaca fuscata*), and India (*Macaca fascicularis umbrosus*) – have recently been seen to use a variety of different materials to floss their teeth.

A group of long-tailed macaques in Lopburi, Thailand are especially likely to use strands of human hair as dental floss (Watanabe, Urasopon, and Malaivijitnond, 2007). These feisty monkeys roam freely around the ancient Buddhist shrine of Prang Sam Yot. Because the local inhabitants worship the monkeys as servants of God, they are given free rein to ride on the heads of human visitors and to brazenly pluck hair from people's heads. The monkeys sort a few strands of hair to form a string with which to floss their teeth by pulling it to and fro between their teeth with both hands. The authors of the report suspect that this practice is a relatively novel cultural behavior, one that has arisen under quite special circumstances because it has not been prominently observed in other groups in nearby locations.

Beyond these interesting observations, the authors conducted a small experiment in which they provided the monkeys with hairpieces. "These monkeys put the hair into their mouths and then pulled it out by grasping the tip with their hands while closing their mouths. When some food became attached to the hair, they licked it off" (941).

These flossing activities were not uniquely associated with hair; the monkeys also used fibers from coconut shells. The animals flossed with these fibrous strands in much the same way as they did with human hair. The tooth flossing observed in India is especially notable for the variety of materials that those monkeys used for food extraction (Pal et al., 2018); a blade of grass, a *Casuarina* plant needle, a feather, a nylon thread, and a metal wire were all used to perform tooth flossing.

Watanabe and company carefully contemplated the behavioral complexity that was involved in the flossing behavior of the Thai monkeys,

> Utilizing women's hair as dental floss is not a simple task; the monkeys need to sort the hair, make a string with it and hold it tightly with both hands to [floss] their teeth when they feel that pieces of food remain. If the hair is not held tightly and/or the gap between the teeth they wish to [floss] is incorrect, they are not able to remove food that is stuck in their teeth. As they are not able to directly see what they are doing, they need to adjust their actions using only their tactile senses. It was interesting that some monkeys appeared to remove only a few pieces of hair as though they understood that there was an optimum number of hairs required for use as dental floss. (943)

The social nature of the flossing behavior of the Thai monkeys – involving several group members of different ages avidly seeking human donors – differs from the largely solitary nature of the flossing behavior detailed in the report on the Japanese monkeys (Leca, Gunst, and Huffman, 2010). Here, only a single "innovator" macaque (Chonpe-69-85-94) was seen to floss her teeth with hair taken from herself or from other group members during grooming (Figure 14.1). However, the practice did not spread to other animals, perhaps because the flossing individual interacted mostly with her mother and younger brother, neither of whom adopted the practice.

TAKING MENTAL FLOSS TO DENTAL FLOSS 157

FIGURE 14.1 Monkey Chonpe-69-85-94 flossing her teeth with her own hair. Author: Jean-Baptiste Leca. Courtesy of Jean-Baptiste Leca.

Nevertheless, the authors of this study did offer some fascinating speculations as to the origins of flossing,

> Because chance may account for a good number of behavioral innovations and dental flossing was always associated with grooming activity, we suggest that the dental flossing innovation is an accidental byproduct of grooming. Thus, the following is a reasonable scenario: during regular grooming episodes, Japanese macaques sometimes bite into hair or pull it through their mouths to remove external parasites, such as louse eggs. Because of ... gaps between incisors, pieces of hair may accidentally have stuck between Chonpe-69-85-94's teeth, and as she drew them out, she may have noticed the presence of food remains attached to them. The immediate reward of licking the food remains off the hair may have encouraged her to repeat the behavior for the same effect in the future, by actively inserting the hair between her teeth. (19)

Of course, the authors were also alert to the more obvious option that, "dental flossing could alleviate the possible physical annoyance caused by a piece of food stuck between the teeth" (19). However, they gave particular weight to the fact that flossing was always associated with grooming as well as to the fact that, in the case of at least this one monkey, flossing was no more likely immediately after eating than at other times during the day.

Leca and company also pondered an even more intriguing possibility in light of Chonpe-69-85-94's tooth flossing over a period of several years. Perhaps a much longer-term beneficial consequence of flossing might have been an improvement in the condition of her teeth and gums. Of course, how those very long-term benefits of prophylactic oral care might have been perceived by the monkey raises extremely challenging issues involving the species generality of foresight and planning (Wasserman, 2019).

This point brings us full circle to the pioneering efforts of Levi Parmly (1819) – the Father of Preventive Dentistry. He proposed that if people were "sufficiently attentive to cleanliness of the mouth, [then] diseases of the teeth and gums might be prevented, without the necessity of any painful operation... [Such care] would contribute no less to the improvement of the features of the countenance, than to the promotion of general health and comfort" (72–73).

A final bit of disconcerting news runs headlong into Parmly's pioneering proposal. Although Parmly firmly supposed that flossing was the most important activity that could be performed with his handy dental kit, his belief – and that of the entire dental profession since Parmly – has recently been called into question.

When the United States government published an update to the 2015–2020 Dietary Guidelines for Americans, it jettisoned the prior recommendation to floss daily. A 2016 Associated Press report contended that officials had inadequately researched the effectiveness of routine tooth flossing; the controversy over the utility of flossing has been dubbed "Flossgate" (Saint Louis, 2016).

At issue here is the fact that very little rigorously controlled research has actually been devoted to assessing the empirical effectiveness of flossing. Why? Many of the studies involved small numbers of subjects over short periods of time. But, perhaps the key problem is that they did not use state-of-the art control procedures,

> The kind of long-term randomized controlled trial needed to properly evaluate flossing is hardly, if ever, conducted because such studies are hard to implement. For one thing, it's unlikely that an Institutional Review Board would approve as ethical a trial in which, for example, people don't floss for three years. It's considered unethical to run randomized controlled trials without genuine uncertainty among experts regarding what works. (Holmes, 2016)

So, is there no longer any need to floss one's teeth? Of course not. The absence of fully vetted evidence in no way undermines the true effectiveness of flossing. Some 200 years of practical experience have convinced the entire dental profession that flossing really does promote dental health. Thus, a 2016 news release from the American Dental Association asserts that, "a lack of strong evidence doesn't equate to a lack of effectiveness. Interdental cleaners such as floss are an essential part of taking care of your teeth and gums. Cleaning between teeth removes plaque that can lead to cavities or gum disease from the areas where a toothbrush can't reach."

Nevertheless, if we humans have determined proper dental hygiene methods by merely suggestive but not definitive evidence, then just how far-fetched would it be that our progenitors and those of today's monkeys may have followed a similar path?

Answering this question will not be easy. However, a comparative perspective may pay considerable dividends in our quest to ascertain the origins of these noteworthy behavioral and technological innovations.

REFERENCES

Chernin, D. and Shklar, G. (2003). Levi Spear Parmly: Father of Dental Hygiene and Children's Dentistry in America. *Journal of the History of Dentistry, 51*, 15–18.

Dominy, S. S., et al. (2019). *Porphyromonas gingivalis* in Alzheimer's Disease Brains: Evidence for Disease Causation and Treatment with Small-Molecule Inhibitors. *Science Advances, 5*, 1–21.

Duenwald, M. (2005, April 21). The Father of Floss. *The New York Times*.

Holmes, J. (2016, November 25). Flossing and the Art of Scientific Investigation. *The New York Times*.

Leca, J. B., Gunst, N., and Huffman, M. A. (2010). The First Case of Dental Flossing by a Japanese Macaque (*Macaca fuscata*): Implications for the Determinants of Behavioral Innovation and the Constraints on Social Transmission. *Primates, 51*, 13–22.

Pal, A., Kumara, H. N., Mishra, P. S., Velankar, A. D., and Singh, M. (2018). Extractive Foraging and Tool-Aided Behaviors in the Wild Nicobar Long-Tailed Macaque (*Macaca fascicularis umbrosus*). *Primates, 59*, 173–183.

Parmly, L. S. (1819). *A Practical Guide to the Management of the Teeth: Comprising a Discovery of the Origin of Caries, or Decay of the Teeth, with Its Prevention and Cure*. Philadelphia: Collins & Croft. https://collections.nlm.nih.gov/catalog/nlm:nlmuid-2566032R-bk

Saint Louis, C. (2016, August 2). Feeling Guilty about Not Flossing? Maybe There's No Need. *The New York Times*. www.nytimes.com/2016/08/03/health/flossing-teeth-cavities.html

Sanoudos, M. and Christen, A. G. (1999). Levi Spear Parmly: The Apostle of Dental Hygiene. *Journal of the History of Dentistry, 47*, 3–6.

Wasserman, E. A. (2019). Precrastination: The Fierce Urgency of Now. *Learning and Behavior, 47*, 7–28.

Watanabe, K., Urasopon, N., and Malaivijitnond, S. (2007). Long-Tailed Macaques Use Human Hair as Dental Floss. *American Journal of Primatology, 69*, 940–944.

FURTHER MATERIAL

American Dental Association News Release (2016, August 4). www.ada.org/en/press-room/news-releases/2016-archive/august/statement-from-the-american-dental-association-about-interdental-cleaners

Be Sure and Floss! Researchers Say Good Dental Health 'Substantially' Decreases Risk of Alzheimer's. (2019, June 5). *Good News Network*. www.goodnewsnetwork.org/good-dental-health-substantially-decreases-alzheimers-risk/

Dental Floss Market Revenue to Rise Substantially Owing to Increasing End-use Adoption. (June 27, 2019). *Financial Planning*. https://financialplanning24.com/dental-floss-market-revenue-to-rise-substantially-owing-to-increasing-end-use-adoption/

Oral Hygiene – History of Dental Hygiene. www.historyofdentistry.net/dentistry-history/history-of-oral-hygiene/

The Family Parmelee. http://thefamilyparmelee.com/x07–0150.html

Primate flossing videos www.youtube.com/watch?v=lHvFFyz2az8&ab_channel=Discovery

www.youtube.com/watch?v=YfiB88WmaII

www.newscientist.com/article/2152868-watch-a-monkey-floss-its-teeth-with-a-bird-feather/

www.bbc.com/news/av/science-environment-45459264/baboons-at-paignton-zoo-have-been-filmed-flossing-their-teeth-with-bristles-and-hairs

15 A Very Close Shave

> Razor in hand, and fully lathered, [the orangutan] was sitting before a looking-glass, attempting the operation of shaving, in which it had no doubt previously watched its master through the key-hole of the closet.
>
> Edgar Allan Poe

These words, from Edgar Allan Poe's famous mystery story *The Murders in the Rue Morgue* serve as prelude to the ape's grisly slaying of Madame L'Espanaye and her daughter Mademoiselle Camille L'Espanaye. This unlikeliest of murderers might have entirely eluded identification if it were not for the astounding perceptual prowess of Poe's shrewd sleuth, C. Auguste Dupin.

Why might Poe have chosen an orangutan to be the killer and for the murder weapon to have been a straightedge razor? As he wrote in the macabre tale itself, Poe stressed this species' well-known "imitative propensities" as one plausible reason. Another more esoteric reason might have been the fact that, among nonhuman apes, orangutans have the most pronounced facial hair surrounding the mouth. Finally, if Poe wanted his killer to be *almost* human, then what better ape could he have chosen than the orangutan? After all, the etymology of the word "orangutan" comes from the Malay and Indonesian terms *orang* (meaning "person") and *hutan* (meaning "forest") – hence, "person of the forest."

But, another even more profound puzzle enshrouds the now common practice of shaving: namely, why do the males of our species go to such trouble to shave off their facial hair? A man's beard represents an especially salient secondary sexual characteristic: a physical feature that develops at puberty, which distinguishes men from women, but which does not directly participate in reproductive acts.

Such sexual dimorphism is widespread in the animal kingdom. Peacocks differ dramatically from peahens as do male lions from lionesses. Charles Darwin famously offered his evolutionary explanation for the development of sexual dimorphism in *The Descent of Man, and Selection in Relation to Sex* (1871) – he called it "sexual selection." Darwin hypothesized that having distinctive male features not only differentiates males from females but it is an effective means for attracting females, thereby allowing females to serve as the ultimate arbiters of male fitness (Oldstone-Moore, 2015). Somewhere in the ancient past, it is believed, women found men with thick facial hair to be preferable as mates than less hirsute men. Beards thus became symbols for reproductive vigor. This preference is especially baffling because, over the course of evolutionary time, we humans lost most of the hair that had grown elsewhere on our bodies in comparison to our primate relations, leading humans often to be called the "naked ape."

Based on Darwin's revolutionary account, a peacock plucking his tail feathers or a lion cropping his mane would be tantamount to committing evolutionary suicide! Thus, deprived of his manifest masculinity, a male could never hope to succeed in competing for a female. Yet, that appears to be exactly what the males of our species do when they remove their beards. Darwin's evolutionary argument thus makes the question, "Why do the males of our species shave off their facial hair?" exceptionally challenging to answer.

To be completely candid, it is unlikely that we'll ever know how the practice of shaving began. Indeed, it is unlikely that we'll ever know whether male humans or males of a human progenitor species began it. Nevertheless, let's concede for the sake of argument that removing facial hair arose as an exclusively human activity. What could have instigated it?

Suppose that the impetus for men's *removing* their beards was the same for men's *growing* them in the first place – women. If women's initial fancy for men with full beards were somehow to change, then women might gradually come to prefer men with more boyish visages. Men's mechanically removing their facial hair might thereby have accelerated what would otherwise have been an excruciatingly slow, evolutionary process.

That account accords with the Darwinian approach, but it nevertheless strikes me as incomplete. What would spur such a profound shift in female choice after such a longstanding preference had been established? I suppose one could buck the current zeitgeist and concur with Virgil who, in The Aeneid, contended that, "a woman is always a fickle, unstable thing." Here is another, less contentious explanation worth considering.

It begins with a simple premise: Having a beard is terribly bothersome. How so? Let's consider some of the ways.

Having myself grown a beard and tended it for twenty consecutive years, I can attest to the fact that beards are difficult to wash and to groom, even with the best available soaps and barbering equipment. Beards not only collect bits of food and dirt but they can also house annoying pests. When the weather gets hot, beards capture the heat and aggravate one's discomfort. When the weather turns frosty, beards can trap moisture; if that moisture were to freeze, then one's skin could more readily fall victim to frostbite. Finally, if men were to engage in hand-to-hand skirmishing, then one's adversary could gain the advantage, as beards are all too easy to grasp. That may have been one reason why, in 331 BCE, Alexander the Great ordered his Greek troops to shave off their beards in preparing for what turned out to be a decisive battle against Persian forces for mastery of Asia.

If it was relief from the many annoyances of beard growth that prompted men to take active measures to remove their beards, then that raises the matter of women's choice between shaven and unshaven men. Here, it seems quite reasonable to propose that, at least for some women, the scruffy and unkempt appearance of bearded men made them less attractive than clean-cut and possibly more youthful looking men. In addition, untamed and prickly beards might prove unpleasant to the touch and chafe women's tender skin.

So, for both men and women, physical comfort might have combined to combat men's previously preferred fuzzy faces. This account does not violate Darwin's explanation for the *origin* of the human beard, but it does suggest that other factors might militate against the *maintenance* of beards.

Although we may never know exactly who initiated the practice of removing beard hair, some interesting insights into the ancient practice of shaving can be gleaned from archeological evidence (Rothschild, 2017; Tarantola, 2014). Based on cave paintings, early humans are thought to have begun plucking their beard hairs approximately 100,000 years ago using hinged seashells as tweezers (Figure 15.1). These men are believed to have used sharpened shark teeth to scrape away annoying beard hair. Some 60,000 years later, slivers of the glasslike volcanic rock obsidian and shards of clam shells were used to slice off beard hairs. By 30,000 BCE, shaving was accomplished with blades of flint that were knapped from larger stones.

Just as there may be more than one way to "skin a cat," there may be more than one way to "remove body hair." Depilatory creams concocted from arsenic, quicklime, and starch began to be used around 3,000 BCE. The Egyptians found these creams to be effective for removing women's body hair, but men needed something much stronger for removing their coarse beard hair. From natural material, pumice stones were used. With the advent of metalworking, metallic blades were fashioned that could yield a much closer shave.

Advances in metallurgy brought about by the Greeks and Romans produced still sharper, smoother razors: first made of bronze and later of copper and iron. The Romans elevated the status of men's shaving by establishing barbershops. These businesses not only provided professional grooming services for their well-to-do clientele but they also emerged as social centers where news and gossip could be unreservedly and ardently exchanged.

Over the centuries, beards have come and gone, much like other fashion trends (Oldstone-Moore, 2015). So too have the various shavers themselves and the accompanying shaving paraphernalia and products, such as shaving cream, aftershave, and skin lotion. Here, special mention goes to one of these shaving products and

FIGURE 15.1 Older man from the Solomon Islands shaving with a clam shell while holding a mirror (1964).

Author: Roger M. Keesing. From Roger M. Keesing Papers (MSS 427). Courtesy of Special Collections and Archives, University of California San Diego especially to the unique advertising materials that were used to promote its sales.

This longtime favorite product is *Burma-Shave*, a brand of brushless shaving cream. I myself never used *Burma-Shave*, so I can't comment on its effectiveness. However, it became famous for the quirky advertising gimmick the company deployed from 1926 to 1963: the posting of amusing rhyming poems on small signs along the side of roadways, with each sign containing one fragment of a longer slogan. So, if you had kids in the car, then as you drove by, the kids would read aloud each of the signs in succession. My brother and I never tired of these recitations, although I suspect that our parents were far less enthusiastic.

Here are a few of the whimsical sayings from the 1920s and 1930s; these warn of the painful consequences that would befall men who failed to use *Burma-Shave*:

> Pity all
> The mighty Caesars
> They pulled
> Each whisker out
> With tweezers

> A shave
> That's real
> No cuts to heal
> A soothing
> Velvet after-feel

> Your beauty, boys
> Is just
> Skin deep
> What skin you've got
> You ought to keep

> Tho stiff
> The beard

> That Nature gave
> It shaves like down
> With
> *Burma-Shave*

Still other sayings attest to the possible advantage of being clean shaven – attracting and keeping women:

> He played
> A sax
> Had no B.O.
> But his whiskers scratched
> So she let him go

> The answer to
> A maiden's
> Prayer
> Is not a chin
> Of stubby hair

> Dewhiskered
> Kisses
> Defrost
> The
> Misses

> Every
> Sheba
> Wants a sheik
> Strong of muscle
> Smooth of cheek

> Before I tried it
> The kisses
> I missed
> But afterward – Boy!
> The misses I kissed

So, it is that, with respect to men's facial hair, "civilization is at war with nature" (Oldstone-Moore, 2015, 5). What millions of years of biological evolution had bequeathed us men, we have proceeded with great resolve and effort to remove. Cultural forces, themselves subject to evolutionary change, can sometimes best the physical results of evolution. For the foreseeable future, men will continue shaving off their beards, only later to regrow them, as tastes cyclically change in the definition of virility to both women and men (Saxton, 2016).

REFERENCES

Oldstone-Moore, C. (2015). *Of Beards and Men: The Revealing History of Facial Hair*. Chicago, IL: University of Chicago Press.

Rothschild, M. (2017, December 29). Things People Used to Shave with before Modern Razors. *Ranker*. www.ranker.com/list/history-of-shaving/mike-rothschild

Saxton, T. (2016, April 14). Hirsutes You Sir: But That Beard May Mean More to Men Than Women. *The Conversation*. https://theconversation.com/hirsutes-you-sir-but-that-beard-might-mean-more-to-men-than-women-56784

Tarantola, A. (2014, March 18). A Nick in Time. How Shaving Evolved over 100,000 Years of History. *Gizmodo*. https://gizmodo.com/a-nick-in-time-how-shaving-evolved-over-100-000-years-1545574268

PART IV Arts, Entertainment, and Culture

16 Ansel Adams
Art for Art's Sake?

The many motion pictures produced by Samuel Goldwyn's studios have prominently portrayed a ferocious lion in its legendary logo. The logo first appeared on the Goldwyn Pictures Corporation silent movie *Polly of the Circus* (1917). It was subsequently used by MGM Studios when Goldwyn Pictures joined with two other film companies in 1924: Metro Pictures and Louis B. Mayer Pictures.

This leonine logo was the product of Howard Dietz, who was later to achieve considerable fame as an advertising executive and songwriter. Dietz chose this specific animal for the studio emblem because a lion was the official mascot of his undergraduate alma mater, Columbia University, as well as the trademark of the school's humor magazine to which he himself contributed. As Dietz wrote in his 1974 autobiography, *Dancing in the Dark*, "I got the idea from the laughing lion decoration in the college comic, *The Jester*. The lion used in the magazine was a symbol of Columbia ... which in turn was taken from the lion on the crest of King's College [Columbia University was originally founded in 1754 as King's College by royal charter of King George II of England]. That's powerful lineage enough for a film company."

Dietz also included in the logo the Latinesque motto *Ars Gratia Artis* on a long looping strip of film surrounding the lion's head (Altman, 1992). Far less was known about the provenance of this motto when, in *The New York Times*, Andresky Fraser (1993) surprisingly claimed that this was a phrase that Dietz "remembered from Latin class."

Her claim was promptly and roundly rebutted, again in *The New York Times*, by Diana Altman (1993), who had just the year earlier published *Hollywood East: Louis B. Mayer and the Origins of the Studio System* (1992). According to Altman, the 19-year-old Dietz had in 1917 recently been hired part-time by the Phillip Goodman Advertising Agency in Manhattan after Dietz won a slogan contest and dropped out of Columbia. Samuel Goldwyn – whose studio was headquartered just across the Hudson River in Fort Lee, New Jersey – retained the Agency in order to produce a full-page advertisement in *The Saturday Evening Post* announcing Goldwyn Pictures to the general public. Dietz was assigned to do the logo work.

In addition to demanding that arresting artwork be featured in the advertisement, Goldwyn insisted that a pithy phrase be added to the logo that would proudly proclaim the studio's resolute commitment to producing, "pictures built upon the strong foundation of intelligence and refinement." Dietz naively assumed that he could directly translate the English phrase "art for art's sake" into Latin and thereby provide the required gravitas to the Goldwyn logo. What could fit the bill better than a classy Latin motto?

After consulting with a Latin scholar, Altman (1993) sought to set the etymological record straight. First, in Latin, the phrase *ars gratia artis* is nonsensical; direct translation fails to capture the sense of art for its own sake. Second, and even worse, the phrase ironically suggests *exactly the opposite* of Dietz's intended meaning,

> In Latin, the word "ars" means art, as in the art of baking bread, a skill or craft. No single word in Latin expressed what we mean by art. In Latin, painting would be "ars picturae," sculpture would be "ars sculpturae." The word "gratia," when used at all, meant doing something for the sake of a goal – just the opposite of how Dietz used it. The concept of art for its own sake was foreign to the Romans. Art for one's patron's sake, maybe.

FINE ART VERSUS PRACTICAL ART

Putting aside Altman's cutting critique of Dietz's ersatz Latin translation, the apprentice publicist had quite properly appreciated that *fine art* might very well be distinguishable from *practical art*. The origin of this important distinction is commonly credited to French author Théophile Gautier. In the introduction to his 1835 novel, *Mademoiselle de Maupin*, Gautier associated fine art with the expression *"l'art pour l'art"* while deprecating practical art with his harsh decree that "everything useful is ugly."

Of course, it was downright audacious for Goldwyn and Dietz to align the primitive and patently commercial movies of their time with fine art, and especially for them to do so on behalf of an upstart motion picture firm. But, the mogul and his apprentice were nothing if not prescient. A mere ten years later at the Ambassador Hotel in Los Angeles – where the center of the film industry had gravitated – the *Academy of Motion Picture Arts and Sciences* was founded on the very idea that this medium was on its inexorable way to evolving into a fine art. Today, the Oscar is among the world's most recognized symbols of fine artistic achievement.

FINE ART VERSUS PRACTICAL ART PHOTOGRAPHY

Photographic artistry can be viewed against the backdrop of motion picture artistry. Although still photographic images greatly predated moving cinematic images, it took considerably longer for photography to attain widespread artistic and creative appreciation. As rich and fascinating as that history is, I'm choosing here to focus on a single famous photographer, Ansel Adams (Figure 16.1), and his own remarkable evolution as an artist; that evolution exquisitely captures the intricate interplay of context, trial-and-error, and chance in human experience. But, before doing so, it will be helpful to distinguish *fine art photography* from *practical art photography*.

Using a camera to document just what appears in front of the photographer falls into the category of photojournalism. Images of this sort seek to capture the reality of the scene at a precise moment in time; such images often serve to elucidate a story. Another form of practical photographic art is meant to promote some commercial enterprise or, less frequently but no less importantly, a political cause.

Fine art photography, on the other hand, is less about the scene than it is about the artist. Fine art photography is not primarily about

FIGURE 16.1 Ansel Adams in 1953.
© Imogen Cunningham Trust

faithfully reflecting what the camera sees; rather, it is about sharing with the viewer what the artist feels. Here, the artist can be said to use the camera as a tool to create a work of fine art in much the same way as a painter uses an easel, brushes, and oils to paint a landscape.

ANSEL ADAMS: FINE ART AND PRACTICAL ART PHOTOGRAPHER

According to the National Archives (archives.gov), which holds a trove of the artist's outstanding work, "Ansel Easton Adams

(1902–1984) was a photographer and environmentalist revered for his black-and-white landscape photographs of the western United States, and his commitment to the conservation of those lands. He helped establish photography as an art form."

What is most important about Adams to us here is that he so laudably flourished in both worlds: the world of practical art and the world of fine art. Most any source you consult will say something like this: Ansel Adams totally transformed the realm of landscape photography. His carefully crafted and stunning black-and-white landscapes opened up nature, providing viewers with an *emotional* rather than a *photorealistic* perspective. Adams proved that nature photography could not only importantly energize and advance conservation efforts but could also successfully compete with other, more traditional genres as fine art.

Adams's own retrospective view of his artistic contributions provides us with only sparse clues to understanding the interrelation between his evolution as a practical art and a fine art photographer. Occasionally, Adams even intimates that there may not be much connection at all. In his autobiography, for instance, Adams (1985) comments that, "People are surprised when I say that I never intentionally made a creative photograph that related directly to an environmental issue, though I am greatly pleased when a picture I have made becomes useful to an important cause." Can there really be no link between these aspects of Adams's art? And, if there is, then just what might it be?

Commonly, biographies richly inform us about an individual's early, formative experiences. However, it turns out that rather little had been written about Adams's early career in photography; that lacuna has been filled by Rebecca Senf's (2020) book, *Making a photographer: The early work of Ansel Adams*. In her invaluable and meticulously documented volume, Senf explores the fascinating interplay between Adams's twin passions of practical photography (promoting commercial clients and environmental causes) and creative photography (evincing fine art). We thus turn to early facets of

Adams's life story, for clues to his development as an effective environmental activist and as an artist of the highest creative order.

ANSEL ADAMS: EARLY INFLUENCES

"My childhood was very much the father to the man I became," wrote Adams (1985). So, indeed, it was. Many of Adams's boyhood experiences profoundly shaped his later life and work. Born in San Francisco on February 20, 1902, the devastating 1906 earthquake permanently and unattractively reshaped 4-year-old Ansel's nasal septum when an aftershock forcefully flung him against the low garden wall of his home. The home was perched on sand dunes scenically overlooking the Golden Gate and Baker Beach, far removed from other homes in the vicinity. The Adams house was quite sturdily built and, for the most part, survived the ravages of the quake thanks to the premium lumber that had been provided by Ansel's prosperous grandfather, who made his fortune as a timber baron.

Adams's father, Charles, had also been a successful businessman; however, in the wake of San Francisco's earthquake and ensuing blaze, his career never fully recovered due to persistently depressed economic conditions. A further challenge for Charles was that Ansel, his only child, was a most unusual boy who required consummate patience and ceaseless attention. Ansel was often ill and moody; yet, he was also extremely active and had great stamina, often roaming far and wide around the wild terrain surrounding the family home. A loner, Ansel didn't socialize with his peers or exhibit suitably respectful comportment toward his schoolteachers. Therefore, Charles was forced to educate his son by himself along with the help of his wife's live-at-home maiden sister and a few special tutors.

Despite these challenging circumstances, father and son forged a relaxed and loving relationship that afforded Ansel wide latitude to explore the world and to discover his own singular place in it. Although the inquisitive Ansel was incessantly flitting from one interest to another, the sheer diversity of these experiences seems to have been vital to his later development as an artist. Some are particularly notable.

Above all else, Adams loved nature. Those many solitary walks did not isolate him from the world; in fact, Ansel constantly communed with the rocks, plants, and animals that he encountered, at one point accumulating a large collection of insects. His love of nature was also joined with an ardent concern over its wanton exploitation. On this particular matter, Adams (1985) in his autobiography relates a poignant story, which can be seen to set the stage for his later interest in environmentalism,

> A beautiful stand of live oaks arched over [Lobos Creek]. In about 2010, the Army Corps of Engineers, for unimaginable reasons, decided to clear out the oaks and brush. My father was out of town when the crime was committed. One of his favorite walks was through these glades to Mountain Lake in the nearby San Francisco Presidio; on his return, he became physically ill when he witnessed the ruthless damage.

As a preteenager, Ansel suddenly and surprisingly exhibited an intense interest in playing the family piano. For many years, he devoted himself to attaining sufficient proficiency to make concertizing a career. At the age of 12, Adams began taking lessons with Marie Butler, whose patient tutelage brought concentration and discipline to his unfocused temperament. Butler also helped Adams appreciate that the fine arts might afford him a means of expressing his deep emotions in a socially suitable fashion. His highly indulgent father later hired other, more senior teachers for Ansel and even purchased an extremely expensive Mason and Hamlin grand piano in 1925 to nurture Ansel's pianistic development. Helping to defray the cost of his own lessons, Ansel gave piano lessons to other pupils (West, 2013).

Many biographers have suggested that Ansel's musical background may have affected his photographic artistry. For instance, Brower (2002) contended that Adams's "approach to photography – his perfectionism, his mastery of tonal scales, the operatic feeling in his grander images – was essentially musical. Adams took photography into a big, moody, exhilarating, Wagnerian country of inky peaks and dazzling snowfields, where no one had climbed before." At least to Adams, music and photography may not at all have been strange bedfellows.

The most critical experience for young Ansel was unquestionably the summer vacation that he and his family took to Yosemite National Park when he was 14. Ansel lobbied hard for the trip after becoming captivated by the majesty of the park through a book, *In the Heart of the Sierras* – a gift from his Aunt Mary during one of his many illnesses. This nearly 400-page volume was written in 1888 by James Mason Hutchings and was richly illustrated with several photographs of Yosemite taken by George Fiske. Adams would later credit Fiske with having been "a top interpretive photographer," much of whose work has been lamentably lost to fires.

Equipped with the Kodak Brownie Box camera that his father gifted him, Ansel took more than 50 snapshots on this first trip, which he later organized in an album. The attraction of the wilderness and the possibility of capturing its natural beauty with photography would, within a dozen years, come to command the remainder of Adams's creative life.

ADAMS'S FORMATIVE YEARS IN PHOTOGRAPHY

That initial visit to Yosemite prompted many more visits over the next several years. Each successive trip meant more snapshots and more experimentation with photographic technique and content by the curious lad. Those highly memorable trips contrasted with the countless hours, days, weeks, months, and years of tedious piano practice. By 1930, photography would finally surpass music as Adams's professional calling.

Many authors accept that Ansel Adams was an entirely self-taught photographic artist and naturalist. Mainly true, that claim fails to credit the valuable assistance he received beginning in 1917 from two Franks: Frank Dittiman and Frank (Francis) Holman (Alinder, 2014; Spaulding, 1998; West, 2013).

Dittiman operated a small photofinishing business in the basement of his home near the Adams's San Francisco residence. Dittiman hired 15-year old Ansel as a part-time "darkroom monkey" for $2.00 a day. Ansel performed numerous odd jobs for Dittiman, such as carrying undeveloped film and finished prints between the shop and local drugstores by streetcar. Ansel also learned how to develop and process photographic negatives. Dittiman thought very highly of Adams's budding darkroom skills, deeming him to be a "natural."

Holman – a retired geologist and amateur ornithologist – served as hiking and climbing guide and mentor on Adams's early trips to Yosemite National Park, helping Ansel carry his gear and directing him to suitable sites to photograph. In time, Adams's mountaineering skills came to best Holman's. Adams later replaced Holman as the keeper or custodian of the Park's LeConte Memorial Lodge during the summers from 1920 to 1924. The lodge served as home base for the Sierra Club, hailed as the most enduring and influential grassroots environmental organization in the nation and currently claiming 3.8 million members and supporters. Adams joined the Sierra Club in 1919 and his first published photographs appeared in the Club's 1922 *Bulletin*.

Adams became increasingly involved in the Sierra Club's social and environmental activities as he shuttled back and forth between living at home in San Francisco and spending summers in Yosemite. He and other Sierra Club members would often take photographs during their yearly extended hiking trips ("High Trips"), later exchanging them with one another. Adams saw the opportunity to aggregate and sell some of his photographs when he returned to San Francisco. At the same time, the amateur artist was advancing his photographic skills, utilizing increasingly sophisticated cameras and devising many of his own unique techniques through hands-on experimentation.

Appreciating that he would have to find some means to support his pursuit of "pure" photography, Adams began seeking out a variety of commercial assignments. He snapped family portraits, photographed homes and buildings for interior designers and architects, and took promotional photographs for sundry businesses including banks and wineries (Senf, 2020). All the while, the commercial Adams was coevolving with the artistic Adams.

On April 17, 1927, Adams took his boldest step in fine-art photography, shooting what he deemed to be his first "consciously visualized" photograph: *Monolith, the Face of Half Dome* [see a reproduction at www.christies.com/img/LotImages/2014/NYR/2014_ NYR_03457_0002_000(ansel_adams_monolith_the_face_of_half_dome_ yosemite_national_park_c_19104106).jpg]. Adams, in 1983, described his pioneering process – from sighting the subject, to snapping the photograph, to developing the print – in the following way, "The ability to anticipate – to see in the mind's eye, so to speak – the final print while viewing the subject makes it possible to apply the numerous controls of the craft in precise ways that contribute to achieving the desired result" (5).

Setting up for the shoot, Adams (1983) visualized the gigantic granite monolith "as a brooding form, with deep shadows and a distant white peak against a dark sky" (4). To help capture his own intense emotional experience – rather than simply snapping a photo-realistic image – meant that Adams had to deploy some unconventional camera settings, including the use of a deep red Wratten No. 29 filter and a long 5-second shutter exposure. Doing so produced a photographic negative that Adams saw "come through" when he removed the negative plate from the fixing bath. Adams subsequently described his exultation with the result, "The desired values were there in their beautiful negative interpretation. This was one of the most exciting moments of my photographic career" (5).

However critical that first photographic step may have been, there were even more "controls of the craft" to be deployed: the negative still had to be developed into a print. For this task, Adams called on some of his freshly devised photofinishing techniques to complete the process. In this connection, Adams highlighted a principal parallel between composition and performance in music and photography, "The negative is comparable to the composer's score and the print to its performance. Each performance differs in subtle ways."

Indeed, benefitting from many more years of experimenting with developing techniques, Adams later made many more prints from his one prized *Monolith* negative, observing that, "my recent prints are far more revealing of mood and substance than are many of my earlier ones" (5). Mood and substance are the very aesthetic qualities that were so obviously lacking in Fiske's own pale and characterless photograph of the same stone megalith; that photograph was entitled, *El Capitan, Half Dome, and Valley*, and was included in Hutchings's (1888) book (30). Far more than Fiske's snapshot, Hutchings's verbal description far better conveys the magnitude and majesty of the Half Dome: "omnipresent as ever, [it] overshadows and eclipses every lesser object" (854).

At this same time, Adams was to benefit from the friendship and patronage of Albert M. Bender, a San Francisco insurance magnate and art aficionado. The day after the two happened to meet at a party, Bender helped Adams prepare, publish, and market his first professional portfolio, *Parmelian Prints of the High Sierras*. Bender's encouragement, business connections, and financial support dramatically changed the course of Adams's life; indeed, that portfolio can be said to have launched his professional career as a photographer. Although the photographs in the *Parmelian Prints* did not display the bold power and striking contrasts of the *Monolith*, they were nevertheless highly appealing to the wealthy private collectors whom Bender knew, thereby helping Adams gain a firmer grasp of marketing; if he was ever going to sell his photographs to prospective buyers, then he was going to have to appreciate their tastes.

Two years later, in 1929, Adams took a job that built on his budding commercial and artistic skills: He was employed as a photographer for the Yosemite Park and Curry Company. As described by Senf (2020), the company sought to attract more tourists to Yosemite. Its marketing department urged Adams to snap eye-catching, appealing photographs. Beyond this general aesthetic directive, Adams was also instructed to capture some very specific scenes in his pictures: In winter, for example, he was to photograph only snow-covered houses and trees as well as to photograph only the best-dressed people using the valley's skating rink. The full gamut of Adams's photographs also included many other vacation activities such as sleigh riding, dog sledding, horseback riding, fishing, golfing, and camping. Not to be completely constrained, Adams also continued to photograph the Park's matchless and vibrant vistas.

These photographs ultimately appeared in a variety of brochures and newspaper articles, on postcards and menus, and in a Yosemite National Park souvenir book. The job afforded Adams many benefits: the chance to further develop his photographic skills, the opportunity to attract more visitors to Yosemite so that they could appreciate its natural beauty, and a good income. Indeed, in 1934, this job paid half of the mortgage on Adams's San Francisco home.

Of course, the aim of the Yosemite Park and Curry Company's advertising campaign was to persuade the public to visit the park. Adams's photographs proved to be especially effective in achieving that aim. The persuasive power of photography to advance the commercial interests of the park was not lost on Adams. After he left the Company in 1937, he later used photography as a powerful tool to advocate for the preservation and expansion of the nation's wilderness areas.

William Turnage, director of The Wilderness Society from 1978 to 1985 and Adams's business manager, has nicely chronicled the photographer's concurrent artistic ascent in the early 1930s (Turnage, 1980a). Adams first visited New York in 1933 to meet Alfred Stieglitz, himself a noted photographer and a prominent promoter of photography as a fine art form. Adams and Stieglitz closely bonded as protégé and mentor, Adams deeply admiring Stieglitz's photographic craft and artistic philosophy.

During the 1930s, Adams spent extensive time in New York, with Stieglitz and others in his photographic circle providing the young artist important opportunities to display his work. In 1933, the Delphic Gallery gave Adams his first New York show. Most significantly, in 1936, Adams was granted a one-person show in Stieglitz's own gallery: An American Place. There, Stieglitz exhibited only those American artists whose work he most favored.

Adams's rising artistic recognition was immensely satisfying; however, it was still not sufficient to meet his escalating monetary obligations. In 1935, Adams wrote to a friend, "I have been busy, but broke. Can't seem to climb over the financial fence." This forced Adams to devote increasing time and energy to commercial photography. His clients now included the National Park Service, Kodak, Zeiss, IBM, AT&T, a small women's college, a dried fruit company, and *Life*, *Fortune*, and *Arizona Highways* magazines. By 1938, Adams wrote to another friend, "I have to do something in the relatively near future to regain the right track in photography. I am literally swamped with 'commercial' work – necessary for practical reasons, but very restraining to my creative work."

Not until later in life – when his fine-art photography had attained national and international prominence, allowing it to be sold at premium prices – did Adams's improving financial situation allow him to escape from the necessity of accepting commercial artwork assignments and to devote himself fully to fine art and environmental advocacy.

ARTISTRY AND ADVOCACY CONVERGE

From 1934 to 1971, Adams served on the Sierra Club's Board of Directors. In his excellent biographical sketch, *Ansel Adams, Photographer*, Turnage (1980b) recounts how Adams leveraged his landscape photography to achieve a key political success.

In 1936, Adams was the Sierra Club's representative at a national and state parks conference in Washington, DC. The Club's leadership selected Adams to present its proposal to create a wilderness park in the Kings River Sierra, believing that a portfolio of his photographs of the region could prove persuasive to Secretary of the Interior Harold L. Ickes, who would also be attending the event. This belief was not founded on idle speculation; photography had proven to be instrumental in making Yosemite Valley a state park in 1864 as well as in making Yellowstone the first national park in 1872. Although the desired park legislation was not enacted that year, Ickes was immensely impressed with Adams's work and invited him to create a photomural for the new Interior Department building.

Ickes later received a copy of Adams's 1938 limited-edition book, *Sierra Nevada: The John Muir Trail*. The Secretary took it to the White House and showed it to President Franklin D. Roosevelt. The President was so enthralled with the beautiful volume that Ickes gave him that copy and asked Adams for a replacement.

Soon after receiving the replacement, Ickes wrote: "My dear Mr. Adams: I am enthusiastic about the book ... which you were so generous as to send me. The pictures are extraordinarily fine and impressive. I hope before this session of Congress adjourns the John Muir National Park in the Kings Canyon area will be a legal fact." Indeed, Kings Canyon National Park did become a reality in 1940 following spirited urging by Ickes and Roosevelt.

Could Adams's book of photographs really have been so politically potent? After the new park had been established, National Park Service Director Arno B. Cammerer confided to Adams, "I realize that a silent but most effective voice in the campaign was your book, *Sierra Nevada: The John Muir Trail*. So long as that book is in existence, it will go on justifying the park." Adams's photographs truly did matter! Not only that, they plainly put the lie to Gautier's earlier decree: everything useful is *not* ugly!

Over the next forty-five years, Ansel Adams experienced both successes and failures as he pursued his fervent environmental agenda, abetted through the medium of photography. Considering the relationship between advocacy and artistry, Adams (1985) perceptively noted that "the approach of the artist and the approach of the environmentalist are fairly close in that both are, to a rather impressive degree, concerned with the 'affirmation of life.'"

Affirmation of the "living earth" was of prime importance to Adams, who put this intensely personal notion into words: "The whole world is, to me, very much 'alive' – all the little growing things, even the rocks. I can't look at a swell bit of grass and earth, for instance, without feeling the essential life – the things going on – within them. The same goes for a mountain, or a bit of the ocean, or a magnificent piece of old wood." This philosophy was one with Adams's photography, "A great photograph is a full expression of what one feels about what is being photographed in the deepest sense and is thereby a true expression of what one feels about life in its entirety."

Given Adams's philosophical reflections, I now see an added meaning that might be given to Brower's (2002) bold claim that, "Of all the great black-and-white photographers, Ansel Adams was the blackest and the whitest." Viewing environmental conservation in black-and-white terms might also be said to hold for Adams's utterly uncompromising fervor to protect the West's wilderness from wanton exploitation and commercialization. Indeed, it proved to be nearly impossible for Adams to control that fervor in his personal and written interactions with President Ronald W. Reagan and his Secretary of the Interior James G. Watt over what Adams believed to be their ruinous environmental policies. Despite all of his best efforts to reverse the course of those policies, Adams ultimately deemed Reagan and Watt to be people who "know the cost of everything and the value of nothing" (Russakoff and Williams, 1983).

Adams genuinely hoped that those entrusted with safeguarding the nation's priceless environmental treasures would have committed themselves to an altogether different vision. In Adams's words: "Let us leave a splendid legacy for our children ... let us turn to them and say, this you inherit: guard it well, for it is far more precious than money ... and once destroyed, nature's beauty cannot be repurchased at any price."

BACK TO THE QUESTION: ART FOR ART'S SAKE?

Hannay (1954) provides some helpful framing for us to consider this question, generally, and with regard to the artistry of Ansel Adams, specifically. Hannay recasts this one question into two parts: what benefits art *alone* affords and what *additional* benefits art may afford.

> To the question "What is the good of art?" meaning what benefits does it procure in terms of something other than art – e.g. does it make people more virtuous or healthy – the quick reply was none, but it is itself good or valuable. Art exists in and for itself, it is intrinsically enjoyable. If you were to enjoy art because it made you healthier, you would really be enjoying being healthier, and the art would be merely a means to this end. (44)

Those who create art may do so for their own gratification or for others' enjoyment. Clearly, young Ansel found taking snapshots of the wilds of Yosemite to be personally pleasurable; so too did his hiking companions on those later Sierra Club outings.

However, when artistic activity is performed for other, unrelated ends, Hannay suggests that we are engaging a different question: namely, "is a work of art productive of other good things apart from itself being enjoyable?" (44–45). This was clearly the case for Adams when he sold his first portfolio, *Parmelian Prints of the High Sierras*, to those well-to-do San Francisco collectors. In this commercial exchange, Adams greatly prized the cash he received; in return, his customers were able to enjoy his photographic art, as it must have satisfied their aesthetic sensitivities.

A more symmetrical exchange took place when Adams was employed by various concerns as a commercial photographer. Here, Adams received much needed remuneration for his photographs; the transaction also furthered the financial interests of his employers – as, for example, when his striking photography boosted the revenue of the Yosemite Park and Curry Company when more tourists visited the park. Do recall that some strings were attached to this deal; the company did specifically prescribe the pictorial content of Adams's work, plainly highlighting the commercial nature of this assignment.

Finally, at the behest of the Sierra Club, Adams deployed his consummate photographic art as a demonstrably effective tool in persuading the federal government to create Kings Canyon National Park. Here, the landscape photographs that Adams had already included in his book *Sierra Nevada: The John Muir Trail* were not expressly taken for this particular political purpose; they were simply aggregated from scenes that Adams had snapped in the course of his own sweeping photographic explorations. Nevertheless, the splendid content and commanding style of those photographs captured the interest and support of officials at the highest levels of government. I suspect that, if Adams were still alive to comment on this specific result of his photographic art, he might very well warm to the catchphrase: *Art for earth's sake!*

The highest public recognition of Ansel Adams's work – as both a photographer and an environmentalist – came when he was awarded the Presidential Medal of Freedom in 1980 by President Jimmy Carter. The president's pronouncement read as follows,

> At one with the power of the American landscape, and renowned for the patient skill and timeless beauty of his work, Adams has been a visionary in his efforts to preserve this country's wild and scenic areas, both on film and on Earth. Drawn to the beauty of nature's monuments, he is regarded by environmentalists as a monument himself, and by photographers as a national institution. It is through his foresight and fortitude that so much of America has been saved for future Americans.

Such wholly deserved, but never sought recognition might well have been the greatest reward of Adams's life.

FINAL REFLECTIONS ON ANSEL ADAMS'S EVOLUTION AS AN ARTIST

Ansel Adams was no prodigy, as he himself seems to have been aware when 14-year-old Ansel ended his note from Yosemite to his father in San Francisco with the seemingly whimsical declaration: "I am your infant prodigy" (Senf, 2020). We cannot help but appreciate the long and meandering path of Adams's personal and professional development: a veritable testimonial to the unmistakable influence of circumstance, consequence, and serendipity. Indeed, given Adams's unusual boyhood and unsettled early adulthood, could anyone – even his own doting and indulging father – ever have predicted the emergence of this masterful photographic artist and staunch environmental activist? Yet, despite the utter improbability of a man as admirable and honorable as Adams ever arising, the world has been immeasurably enriched by his life and work.

How then might we most fittingly conclude our discussion of Adams's evolution as a fine-art photographer? Here's just a bit of what he himself said about what went into making a photograph, "You don't take a photograph, you make it. You don't make a photograph just with a camera. The single most important component of a camera is the twelve inches behind it. You bring to the act of photography all the pictures you have seen, the books you have read, the music you have heard, the people you have loved." You may wish to recall these thoughts as well as a final observation when you next view one of Adams's spectacular photographs, "There are always two people in every picture: the photographer and the viewer." You will not be alone – Ansel Adams will be there with you!

REFERENCES

Adams, A. (1983). *Examples: The Making of 40 Photographs*. New York: Little, Brown and Company.

Adams, A. (1985). *Ansel Adams: An Autobiography*. New York: Little, Brown and Company.

Alinder, M. S. (2014). *Ansel Adams: A Biography*. New York: Bloomsbury.

Altman, D. (1992). *Hollywood East: Louis B. Mayer and the Origins of the Studio System*. New York: Tapley Cove Press.

Altman, D. (1993, February 7). STUDIO LOGOS; Leo the Lion Flunked Latin. *The New York Times*. www.nytimes.com/1993/02/07/arts/l-studio-logos-leo-the-lion-flunked-latin-815193.html

Andresky Fraser, J. (1993, January 17). What's in a Symbol? Not the Statue of Liberty. *The New York Times*. www.nytimes.com/1993/01/17/archives/film-whats-in-a-symbol-not-the-statue-of-liberty.html

Brower, K. (2002, June/July). Ansel Adams at 100. *The Atlantic*. www.theatlantic.com/magazine/archive/2002/07/ansel-adams-at-100/302533/

Dietz, H. (1974). *Dancing in the Dark: Words by Howard Dietz: An Autobiography*. New York: Quadrangle-New York Times.

Hannay, A. H. (1954). The Concept of Art for Art's Sake. *Philosophy, 29*, 44–53.

Hutchings, J. M. (1888). *In the Heart of the Sierras*. Oakland, CA: Pacific Press Publishing House. www.yosemite.ca.us/library/in_the_heart_of_the_sierras/in_the_heart_of_the_sierras.pdf

Russakoff, D. and Williams, J. (1983, July 3). The Critique. *The Washington Post*. www.washingtonpost.com/archive/politics/1983/07/03/the-critique/32064a44-add8-4ea1-8564-3c7eb09e79cb/

Senf, R. A. (2020). *Making a Photographer: The Early Work of Ansel Adams*. New Haven, CT: Yale University Press.

Spaulding, J. (1998). *Ansel Adams and the American Landscape: A Biography*. Berkeley, CA: University of California Press.

Turnage, W. (1980a). *Ansel Adams, Photographer*. www.anseladams.com/ansel-adams-bio/

Turnage, W. (1980b). *Ansel Adams: The Role of the Artist in the Environmental Movement*. www.anseladams.com/environmentalist/

West, K. (2013). *Ansel Adams*. New York: Chelsea House.

FURTHER MATERIAL

Howard Dietz. Songwriters Hall of Fame. https://web.archive.org/web/20160826195840/http://www.songwritershalloffame.org/index.php/exhibits/bio/C62

Oden, L. Ansel Adams | International Photography Hall of Fame. https://iphf.org/inductees/ansel-adams/

The Scribner Encyclopedia of American Lives (1998). *Volume One, 1981–1985. Howard Dietz Biography*, 233–234. New York: Charles Scribners.

17 Basil Twist: "Genius" Puppeteer

> Puppetry is an abstraction of the spark of life
>
> Basil Twist

"Say Kids! What Time Is It?" For me and for millions of other American Baby Boom children born between 1946 and 1964, it was time for *The Howdy Doody Show*. Beyond the many improbable characters that populated Doodyville – including Buffalo Bob, Princess Summerfall Winterspring, Clarabell, and Chief Thunderthud – the freckle-faced string-puppet Howdy Doody was the undeniable star of the show.

But, the origins of puppetry stretch back much, much farther than that iconic television program. As theater, puppetry has existed for as long as 4,000 years and has been traced to Europe, Asia, and Africa. Some historians believe that puppet theater and human theater share a common beginning and may even have developed in tandem. Rather than representing a simpler performance form than human theater, puppet theater has been deemed to be more complicated, less direct, and more time consuming and effortful to create. Even now, puppetry continues to evolve – from representational to abstract and from traditional to avant-garde.

One of today's leading puppeteers is Basil Twist (see photographs and videos of Twist in his New York workshop at www.macfound.org/fellows/949/#photos), who excels in both traditional and avant-garde performance. I first learned of Twist through an article in *The New Yorker* (Acocella, 2013). I later attended his exceedingly engaging and informative presentation at The University of Iowa on September 29, 2016, which amply demonstrated his consummate talents. I've continued to follow Twist's exploits, which included his creating the puppets used in the Joffrey Ballet's 2016 *Nutcracker* that debuted in Iowa City at Hancher Auditorium.

Twist vaulted to fame in 1998 largely because of his breakthrough performance of *Symphonie Fantastique*. That work propelled his receipt of a 2015 MacArthur Foundation Fellowship (a no-strings-attached award of $625,000); he also received other accolades, including a Guggenheim fellowship, as well as Obie, Drama Desk, Bessie, and Creative Capital awards.

What do we know about Basil Twist's background and his creation of *Symphonie Fantastique*? Several recent interviews have allowed him to tell his own history (Acocella, 2013, 2018; Harss, 2018; Raymond, 2018; Tilley, 2016). In what follows, I've used direct quotations assembled from all of these interviews.

Born in 1969, Basil Twist is a native of San Francisco, California. His mother and grandmother were both puppeteers. At the age of three, his parents gave him his first puppet theater. Basil began making puppets for his theater out of paper. When Basil was ten, his father built a wooden puppet theater for his more elaborate productions. At this time, Basil also inherited a collection of realistic puppets characterizing several big-band jazz stars; these allowed him to cultivate his manual puppeteering skills. Basil later became engrossed in the puppets from *Sesame Street* and the *Muppet Show*. When *Star Wars* was released in 1977, Basil constructed several puppets of the cast.

After high school, Twist briefly attended Oberlin College. The small school and quaint midwestern town were not to his liking, so he moved to New York City. Twist enrolled in a few courses at NYU and began to gain more personal contact with others in his profession. There, he heard about a unique school where he could advance his proficiencies to even higher levels.

Twist applied to and was accepted by the École Supérieure Nationale des Arts de la Marionnette, an international school of puppetry located in Charleville-Mezieres, France. Its international reputation was well earned. "The experience of being in a French school, learning from a Brazilian master of an American technique and then supported by a Swedish master who taught me a Malaysian style to control this puppet is, in a nutshell, the privilege I had being in this world of puppetry and particularly this school," he said. Incidentally, Twist is the only American to have graduated from the prestigious school.

At the École, Twist was introduced to representational puppetry. There, the French troupe named the Ballets Russes performed an abstract ballet without dancers which captured his imagination. Many years later, he would find fame in the same modern genre.

Returning to New York City, Twist spent the next five years performing in nightclubs as well as for friends and acquaintances. He also created a small one-person show of his own that was part of the Henson International Festival of Puppet Theater. Twist was very pleased with his career at this point, "I got on the cover of *The Puppetry Journal*. I thought that it was about as far one can go and that I had made it!" But, then serendipity struck!

One day in 1995, while Twist was out for a stroll and scouring through recycling piles for performance materials, he spotted an aquarium with a small crack in it. "'Hmm,' I said, and I lugged it home, and I covered the crack with duct tape, and I filled the thing with water. I put a piece of silk on the end of a coat hanger, and I dragged it through the water. It looked fantastic."

Flush with $1,000 from his first grant from the Henson Foundation, Twist purchased the largest fish tank he could afford and experimented with a variety of other items put into it. "Almost everything, when you put it under water, is transformed and looks great. You put a black garbage bag under water, it looks great. Cotton balls, not so good. But feathers, great." Further impromptu experimentation involved more intricate choreography, including plunges, glides, turns, and – it almost goes without saying – twists!

Twist next tried setting this kinetic ballet to music. He recalled one of his earlier European experiences, "When I was in France, there was a festival of puppetry and music. It was provocation to explore the relationships between puppetry and music and it immediately made me think that I want to see something abstract relating to music." But, what piece of music would be best for the performance?

Once more, serendipity struck! "Again, it was something from the street. I was walking past a record shop that had milk crates on the sidewalk and I saw an album of *Symphonie Fantastique* with this weird psychedelic cover. It had the image of [a] sunflower with two faces. So, I bought the record. I also remembered the title from my childhood because my parents had it in their record collection. I listened to the record and then I had these fantastic dreams."

Subtitled "An Episode in the Life of an Artist, in Five Parts," Berlioz's revelatory 1830 work is programmatic, telling the surrealistic story of an artist in opium-infused despair over his unrequited love for a beautiful but elusive actress. Famed conductor and composer Leonard Bernstein called this the first musical trip into psychedelia!

Twist proceeded to explore its possibilities for his performance. "The piece has five movements and I had this really ambitious idea that I was going to do each movement with a different element – smoke, fire. I'd already played with water, so I decided I was going to do the third movement underwater. ... But [the aquarium] was so heavy and hard to move that I decided to do the whole piece underwater. The main thing was the idea of abstraction, and then the water thing was just a cool way to achieve that. It was just very carefree the way the whole piece came about."

Staging the show did not prove to be so carefree; considerable physical effort, trial-and-error experimentation, and pragmatic innovation were needed. During performances, water splashes everywhere around the 1,000-gallon tank, with about 100 gallons lost per show. A platform had to be built around the tank to serve as a staging area. Five puppeteers in wetsuits had to straddle the sides of the tank or slip into harnesses, so that they could hover above the water, manipulating wires affixed to fabric, bubbles, feathers, film, pinwheels, and tinsel (Figure 17.1). Then, the puppeteers could drop these various props into the water below at predetermined times in the musical score.

Of course, the musical accompaniment also had to be integrated into the performance. Twist's original 1998 show used a recording of

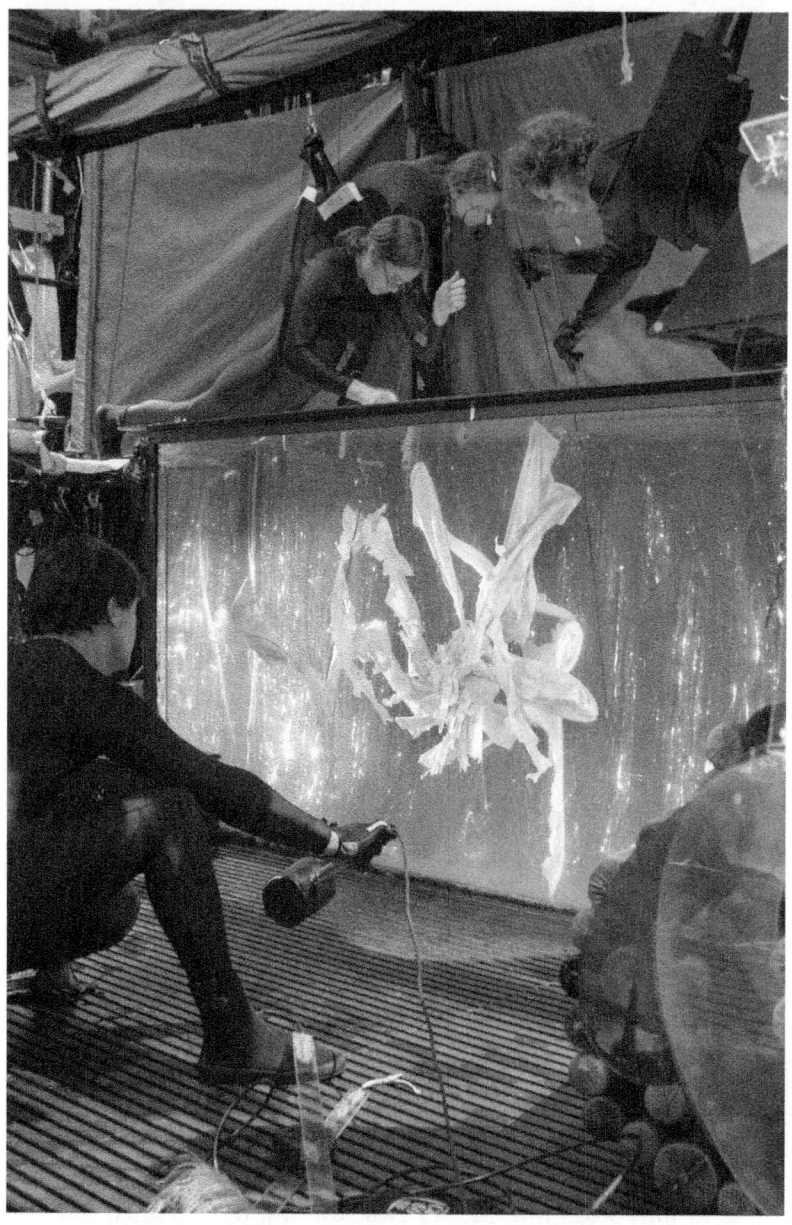

FIGURE 17.1 Basil Twist's team of backstage puppeteers in wetsuits rehearsing "Symphonie Fantastique".
Author: Richard Termine. Courtesy of Richard Termine

Hector Berlioz's symphonic score. The 2018 revival used a live pianist: Christopher O'Riley played Franz Liszt's remarkable arrangement of the *Symphonie Fantastique* – an arrangement actually done in close collaboration with Berlioz.

In a 2018 interview in *The New York Times*, Twist was still stunned that this amazing assemblage of bits and pieces worked together as well as it did. Surveying the drenched props and the puddles on the floor after a show, he wondered aloud, "wow, how did I do that?"

Here, we can ask a related question: namely, was this extremely engaging and creative work a product of prescient planning? One effusively enthusiastic reviewer (Brantley, 2018) suggests just such an account, "Mr. Twist's ballet is a reminder that what we call ineffable art may be a product of meticulous planning and execution."

From what Twist himself has reported, a very different account can be offered that deemphasizes foresighted planning; it instead stresses chance, unpremeditated experimentation, and diligent work. From a broken fish tank to an innovative pièce de résistance, we see a kind of Darwinian variation and selection in action – with absolutely no end in sight. This account in no way diminishes Basil Twist's inventive handiwork. It does, however, place it in a very different light – one that questions the common penchant to anoint exceptionally accomplished people like Basil Twist as "geniuses."

REFERENCES

Acocella, J. (2013, April 8). Puppet Love. *The New Yorker*. www.newyorker.com/magazine/2013/04/15/puppet-love

Acocella, J. (2018, March 23). The Return of Basil Twist's Underwater Puppet Show. *The New Yorker*. www.newyorker.com/magazine/2018/04/02/the-return-of-basil-twists-underwater-puppet-show

Brantley, B. (2018, April 4). Review: Head Tripping the Light Fantastic in 'Symphonie Fantastique.' *The New York Times*. www.nytimes.com/2018/04/04/theater/symphonie-fantastique-review-basil-twist.html

Harss, M. (2018, March 23). Basil Twist's Fantastic Feathered World (with Tinsel and Berlioz). *The New York Times*. www.nytimes.com/2018/03/23/arts/dance/basil-twist-symphonie-fantastique-here.html

Raymond, G. (2018, April 24). Revisiting the Past: An Interview with Master Puppeteer Basil Twist. *Slant*. www.slantmagazine.com/interviews/revisiting-the-past-an-interview-with-master-puppeteer-basil-twist/

Tilley, R. (2016). Basil Twist Uses Puppetry to Explore the Line between the Living and the Inanimate.https://research.uiowa.edu/impact/news/basil-twist-uses-puppetry-explore-line-between-living-and-inanimate

FURTHER MATERIAL

Symphonie Fantastique Trailer https://vimeo.com/248033993

Master Puppeteer Basil Twist Revisits His 1998 Hit Show www.timeout.com/newyork/theater/symphonie-fantastique

Puppeteer Basil Twist at His Studio in the West Village www.youtube.com/watch?time_continue=198&v=n1YatoKv3yU

Puppetry Artist and Director Basil Twist, 2015 MacArthur Fellow www.youtube.com/watch?v=TE9LBBrVaTw

18 Moonwalking: And More Mundane Modes of Moving

1969: Neil Armstrong walks on the moon.
1983: Michael Jackson moonwalks on the earth.

At first blush, these two memorable moments in human history appear to have nothing in common, except for the fact that the term "moonwalk" commonly denotes each. But, consider what makes these two celebrated achievements so salient. Neil Armstrong defied gravity by being hurtled into space by a powerful Saturn V rocket, by landing on the moon in the Lunar Excursion Module (LEM), and by leaping from the ladder of the LEM onto the powdery lunar surface to take his brief first stroll. Michael Jackson appeared to defy the laws of gravity by smoothly sliding backward while stepping forward in his signature hip hop dance move.

For casual observers, these short synopses suffice; but, for those who are interested in how behaviors come to be, one must delve deeper. Discovering the origins of behavior – whether renowned or unremarkable – requires more careful consideration. The origin of walking proves to be a revealing realm to explore, whether on the moon or on the earth.

LUNAR LUMBERING

Rising from the gravitational attraction of earth was a longstanding dream of humanity. Heavier-than-air flight proved that technology could at least temporarily overcome gravity. But, breaking free from the earth's gravitational pull was only to be achieved much later. Most famously, the Apollo program – which ran from 1961 to 1975 – succeeded in placing a dozen men on the moon in six lunar landings (Apollo Program Summary Report, 1975). Even more daring was having the Apollo astronauts actually walk on the lunar surface.

FIGURE 18.1 Astronaut Buzz Aldrin walks on the surface of the moon near the leg of the lunar module Eagle during the Apollo 11 mission. Mission commander Neil Armstrong took this photograph with a 70 mm lunar surface camera.
Courtesy of National Aeronautics and Space Administration (NASA)

Neil Armstrong and Buzz Aldrin (Figure 18.1) did just that five decades ago on July 20, 1969.

This amazing achievement was not casually undertaken. In preparation for this unprecedented exploit, the National Aeronautics and Space Administration (NASA) devised highly specialized training devices and techniques to simulate the weak gravitational pull of the moon (Shayler, 2004). NASA even transported the astronauts to remote and inhospitable earthly venues to emulate the lunar terrain on which the space voyagers would have to walk.

What were NASA's devices and techniques? What were they envisioned to accomplish? First and foremost was having the crew members learn to walk and to remain upright under simulated conditions of one-sixth gravity. The reduced gravity chair (technically termed the Partial Gravity Simulator and whimsically dubbed *POGO* after the pogo stick, a familiar toy used for hopping up and down by means of a spring) was an ingenious contraption that decreased the earth's normal pull by means of a complex concatenation of cables and booms (Figure 18.2). Another method involved parabolic up-and-down flights, which were taken in KC135 airplanes (the transport version of the Boeing 707 airliner); the flights dropped the earth's gravitational draw to one-sixth of normal for about 30 seconds on each descent. Finally, the astronauts practiced walking in a water tank wearing extra weights to give them an effective weight of one-sixth gravity; this aquatic training also allowed the astronauts to practice lifting and maneuvering bulky scientific equipment, which was a key element of their lunar duties.

A further issue revolved around the geological conditions that would be encountered on the moon. Here, the moon's dusty, crater-pocked surface was simulated by having the astronauts train at a barren and rocky desert locale – Meteor Crater – outside of Flagstaff, Arizona.

Of course, all of those preparations were only approximations of the real circumstances that would be experienced on the moon itself. Once on the lunar surface, the astronauts showed amazingly rapid learning to adjust to the dramatic decrease in weight that they felt. Neil Armstrong reported that, "there seems to be no difficulty in moving around – as we suspected. It's even perhaps easier than the simulations of one-sixth g that we performed in the various simulations on the ground. It's absolutely no trouble to walk around."

Armstrong's partner, Aldrin, made additional useful observations. He reported feeling "buoyant" on the moon's surface. On the

FIGURE 18.2 The Partial Gravity Simulator, or POGO. The POGO supports 5/6th of the astronaut's weight in order to simulate walking on the lunar surface.
Courtesy of National Aeronautics and Space Administration (NASA)

earth, the backpack and suit weighed 360 pounds, but on the moon it weighed only 60 pounds. Nevertheless, the pressurized and thermally controlled suits that the astronauts donned were as "hard as a football" and they hampered most strenuous activities; in particular, bending over was extremely difficult. As well, the backpack altered the astronauts' center of gravity; they felt balanced only when they were tilted slightly forward.

NASA recordings depict Aldrin moving about the moon to test his maneuverability. He reported some awkward sensations going forward and then stopping. "You have to be rather careful to keep track of where your center of mass is. Sometimes it takes about two or three paces to make sure you've got your feet underneath you. About two to three or maybe four easy paces can bring you to a fairly smooth stop." Shifting direction posed another challenge. "To change directions, like a football player, you have to put a foot out to the side and cut a little bit."

Aldrin also experimented with two other methods of moving about. He tried two-legged hops, leaping forward with both legs together. "The so-called kangaroo hop does work, but it seems as though your forward mobility is not quite as good as it is in the more conventional one foot after another." He preferred a "lazy lope" that covered about a yard with each long stride, floating with both feet in the air most of the time in a style that was reminiscent of a football player careening through a gauntlet of old tires in a training drill. However, even that preferred method proved rather tiring after several hundred steps. He suspected that this fatigue "may be a function of this suit, as well as the lack of gravity forces."

Regardless of that uncertainty, over the course of the remaining five moon landings, later astronauts were able to build on the initial experiences of Armstrong and Aldrin, and to travel farther, faster, and more sure-footedly. They could also relax a bit more and better enjoy the experience of moon walking. Scuttling down a slight incline, Apollo 16 crew member John Young exclaimed with delight, "Boy, is this fun!" And, Gene Cernan and Jack Schmitt exchanged these remarks while "singing and strolling" on the moon during the Apollo 17 mission, "Boy, this is a neat way to travel." "Isn't this great?" Yes, it was! And, it was the result of both extensive preparation and the direct experiences of the astronauts themselves.

BACK TO THE BACKSLIDE

Ask a hundred people to name the dancer who invented the moonwalk and virtually all of them will unhesitatingly answer, "Michael Jackson." The moonwalk was certainly Michael Jackson's signature dance move. He is said to have chosen that name for the move because of its unearthly effect of forward steps mysteriously moving the dancer backward.

Jackson's first public performance of the moonwalk was on March 25, 1983. But, others had performed that dance or its precursors far earlier. But, how much earlier? Video records are available to help answer this question.

Cab Calloway performed a series of backsliding moves – what he and others called the "Buzz" – among many other twisting gyrations during the opening credits of the 1932 short cartoon entitled *Minnie the Moocher*, a Fleischer Studios Talkartoon, which starred Jazz Age flapper Betty Boop and her dog Bimbo. Tap dancing sensation Bill Bailey also performed striking backsliding maneuvers as he moved backward during his stage exit in 1955. Mahmud Esambaev, a Chechen dancer, performed a most urbane backslide in white tux and tails in 1962. In a similar fashion, while he was effectively standing still, rock star David Bowie adroitly took clear forward steps as he seemingly slid backward in a 1974 concert performance of his album *Aladdin Sane*.

Most notably, Los Angeles singer-songwriter and dancer-choreographer Jeffrey Daniel may lay claim to having first performed the dance move that we now know as the moonwalk. This moment is documented in a 1982 video from a British Broadcasting Corporation *Top of the Pops* broadcast, in which Daniel danced to the song that he had earlier recorded with his Rhythm and Blues group Shalamar: *A Night to Remember*. That video proves that Daniel can be credited with joining the backslide dance move with compelling forward-moving mime.

Daniel's performance was a huge hit. Indeed, that one appearance prompted a return engagement on the BBC. But, the Brits were not the only ones to enjoy that show. Michael Jackson also viewed Daniel's stunning dance routine and later asked Daniel to teach it to him. Jackson proved to be a quick study and, after mastering Daniel's demanding maneuvers, gave the dance its famed name – the "moonwalk."

All of this history is extremely interesting; but, does it allow us to trace a simple linear path to March 25, 1983? Hardly! There are clear gaps in the historical record much like those cases of "missing links" in the evolution of our species. Those gaps raise questions and provoke spirited debate among those trying to write a coherent narrative on the moonwalk.

Note that this evolutionary debate concerns a period of only fifty years with written records and video archives available to the researcher. Now, imagine the extraordinary challenge of tracing the origins of human behavior over millions of years with only fragmentary paleontological evidence at our disposal. That captures the very meaning of "daunting!"

WALK LIKE A MAN

Of course, along with the spoken word, *bipedal* locomotion tops the list of purportedly unique human abilities. But, just how unique is two-legged walking among the great apes? And, is it predestined that all humans and only humans walk upright? Several observations prompt us to give these questions greater scrutiny.

Ambam is a 30-year-old male silverback western lowland gorilla who resides at Port Lympne Animal Park in Kent, England. He became an instant interspecies sensation when video was posted on the internet showing him strolling upright around his grassy enclosure. Most experts agree that occasional, brief bouts of bipedal locomotion are not unknown among many primate species, including bonobos, chimpanzees, and gorillas; however, such locomotion does not occur more that 5 percent of the time and then usually in the context of holding objects in one or both hands. Therefore, taking a long series of continuous steps, as does Ambam, is quite rare.

What are we to make of this rarity? It is tempting to speculate that human bipedality did not arise de novo, but gradually emerged over extended evolutionary time. Bipedal locomotion may thus have originated earlier, millions of years ago. It may have been present in an ancestral species that is common to humans and the African apes (Deaton, 2019) or it may have evolved independently in other prehuman apes (Bower, 2019). We simply don't know.

Of additional importance is the development of bipedal walking in the lifetime of individual organisms. In this connection, Tracy Kivell of the Max Planck Institute for Evolutionary Anthropology has observed that infant and juvenile gorillas engage in many different types of locomotion as they are maturing. These young apes seem to be learning how to move about their environs as they keep pace with other more senior individuals in the troop.

The role of experience may be particularly important in the case of Ambam (Hull, 2011). When he was a year old, Ambam fell ill and had to be taken from his biological mother. Zookeeper Jo Wheatley was then recruited to serve as Ambam's foster mother for a year. During that time, Ambam learned to eat off plates and to drink from a cup. Wheatley recalls that Ambam was standing and walking even then, "from the outset he was a very unusual gorilla. He has always been adept at standing on his hind legs. He often preferred to be fed standing up, sucking on a bottle."

Of course, growing up in such close company with humans means that Ambam could have modeled his walking behavior after Wheatley and other people whom he saw. As well, Ambam may have engaged in far more manual activities than most normally reared gorillas. Wheatley recalls that Ambam, "would often sit down and watch television with us at night and [he] learned how to turn it on and off himself." In another circumstance, Ambam was spared the indignity of being bathed by his human companions, "but if one of us was in the tub he would come and stand up on his hind legs alongside and splash and bat the water."

Such extensive and extraordinary cross-fostering might not be necessary to encourage upright walking, however. Another male western lowland gorilla, Louis, also walks upright. He was born at the St. Louis Zoo in 1999 and later moved to the Philadelphia Zoo in 2004 along with his gorilla parents. Louis had none of the extremely close experiences with humans as had Ambam. "Gorillas will occasionally stand up for a few seconds or walk a couple of steps, but what we see Louis doing is really walking clear across the yard, and that's quite special," said Michael Stern, a curator at the Philadelphia Zoo. Nevertheless, one cannot dismiss the possibility that Louis might have been imitating the walking of zoo staff or spectators.

How unusual then are the behaviors of these two captive gorillas? Perhaps not so special. The wild chimpanzees of Bossou Forest located in southeastern Guinea, West Africa are well known for their prodigious use of a stone hammer and anvil to crack open nuts. The chimpanzees are also frequent bipedal walkers. This mode of locomotion is particularly prominent when rare, highly valued nuts are made available. The chimpanzees collect the nuts and, together with the stone hammer and anvil, move them to a more secluded place where they can eat in peace without having to share.

The authors of this study (Carvalho et al., 2012) deem these observations to support a prominent theory of human bipedalism: namely, that deforestation forced our ancestors to forage in more open settings, thereby requiring them to carry food over greater distances (Hewes, 1961). Beyond these evolutionary speculations, it is now clear that at least two of our closest evolutionary relatives – gorillas and chimpanzees – are quite capable of walking upright.

And, speaking of close relatives, it turns out that several closely related humans in Turkey have adopted a most unusual *quadrupedal* method of locomotion. These five Kurdish brothers and sisters walk on the palms of their hands and feet. Some researchers have speculated that this anomaly represents a backward step in evolution (so-called devolution), with these individuals moving about in a manner possibly more typical of nonhuman primates such as monkeys and apes.

This conjecture turns out to be incorrect. The five Kurdish siblings suffer from Uner Tan Syndrome, a neurological condition in which the cerebellum is underdeveloped, leading to a loss of balance and coordination. Painstaking research involving detailed video records has actually revealed that the precise quadrupedal gait of these individuals is not at all like that of nonhuman primates; rather, the quadrupedal gate of these individuals is just like that of healthy adult humans instructed to locomote on all fours in a controlled experimental situation (Shapiro et al., 2014). Thus, people with Uner Tan Syndrome regularly walk on all fours because of biomechanical constraints of human anatomy not because this specific form of quadrupedalism is an evolutionary throwback.

How individuals with Uner Tan Syndrome come to walk on all fours raises intriguing, but as yet unanswered, developmental questions. On the more general matter of locomotor development, we actually know surprisingly little about how normal humans learn to walk. But, we are discovering more and more thanks to ongoing lines of empirical research.

TAKING BABY STEPS

How babies learn to walk is an extremely interesting and intricate process (see review by Adolph and Franchak, 2016). Both in the womb and soon after birth, infants perform precursory stepping movements. Notably, newborns move their legs in an alternating pattern if they are held upright with their feet on a solid surface, as if they are walking. These initial movements are later shaped by practice, leading to strengthening of the leg muscles and to increasingly adaptive performance.

Of course, human infants' progress toward adult walking is anything but a straight-line, pain-free process. By the end of their first year, babies will likely have experimented with many different means of moving about, including rolling like a log, shuffling on the buttocks, crawling on the stomach or on hands and knees, and cruising sideways along walls, furniture, and objects. Bumps, bruises, and cuts often mark the progress of this decidedly trial-and-error process.

Researchers have for 100 years sought to classify these different kinds of movements and tried to assign them to fixed chronological ages. Those researchers have further proposed that locomotor development represents an innately programmed progression of stages that inexorably unfold due simply to the child's neural-muscular maturation. Recent research undermines this fixed maturational interpretation. Rather, babies must *as individuals* learn to move their bodies from place to place, with this learning critically depending on their growing bodies and their earlier modes of moving.

Babies do indeed learn to walk within a rather short time, taking their first walking steps around twelve months of age (within a range of eight to eighteen months). Nevertheless, in that time, there is ample opportunity for them to perfect the skill of locomotion. For example, in a single hour of playtime, a fourteen-month-old toddler takes some 2,400 steps, covers a distance of eight football fields, and takes some seventeen falls.

Further complicating this story is the fact that there are important cultural factors influencing the trajectory of babies' learning to walk (Gupta, 2019). For instance, infants in the Central Asian country of Tajikistan are traditionally cradled in wooden gahvoras, in which babies are swaddled in long swaths of fabric permitting only their head to move. Such constricted movement for much of the first two or three years of life might be expected to have grave effects on babies' later locomotor behavior. Yet, when finally freed from this protective confinement, these children show no obvious deficits in their walking, running, or jumping. Clearly, we need to widen the scope of our developmental studies if we are to make sweeping generalizations concerning human locomotor behavior.

Beyond learning how to walk, babies devote much of their time and energy to physically interacting with available objects, thereby learning how to move their hands and arms. Precious little of that interaction and learning is tightly scripted; it cannot be, because there is no way for any of us to be prepared for the precise circumstances to which we will have to adapt. Variability and novelty are the rule, not the exception. Thus, adapting our movements to the prevailing environmental constraints demands a continual process of online problem solving that must be improvisational and flexible.

Changes in bodies, skills, and environments are particularly dramatic during infancy, but even as adults we never cease developing. We are always learning to move. Early development is actually an exaggerated version of the temporary changes that result from the flux of everyday life. Environments and bodies are always variable. A gravel road or high-heeled shoes change the biomechanical constraints on balance and locomotion. A sprained ankle or a blister on the foot change the force and span of our gait. Adaptive responding to the unstable conditions and varying vicissitudes that are characteristic of everyday life is something we all do. It is also what the "moonwalking" astronauts and dancers did. Experimenting with physical limits and making mistakes are inherent parts of successful adaptation. Little can truly be taken for granted, even with the most practiced of behaviors – like taking an afternoon stroll.

REFERENCES

Adolph, K. E. and Franchak, J. M. (2016). The Development of Motor Behavior. *WIREs Cognitive Science, 8*, 1–18.

Apollo Program Summary Report (JSC-09423). (1975). National Aeronautics and Space Administration, Lyndon B. Johnson Space Center, Houston, TX.

Bower, B. (2019, November 6). Fossils Suggest Tree-Dwelling Apes Walked Upright Long before Hominids Did. *Science News*. www.sciencenews.org/article/fossils-suggest-tree-dwelling-apes-walked-upright-long-before-hominids-did

Carvalho, S., Biro, D., Cunha, E., Hockings, K., McGrew, W. C., Richmond, B. G., and Matsuzawa, T. (2012). Chimpanzee Carrying Behaviour and the Origins of Human Bipedality. *Current Biology, 22*, R180–R181.

Deaton, J. (2019, September 30). Ancient Ape Fossil Yields Surprising New Insights about Human Evolution, *NBC News*. www.nbcnews.com/mach/science/ancient-ape-fossil-yields-surprising-new-insights-about-human-evolution-ncna1055916

Gupta, S. (2019, September 10). Culture Helps Shape When Babies Learn to Walk. *Science News*.

Hewes, G. W. (1961). Food Transport and the Origins of Hominid Bipedalism. *American Anthropologist, 63*, 687–710.

Hull, L. (2011, January 29). Monkeying Around: We Tell the Story of How Ambam the Walking Gorilla Took His First Steps to Global Fame. *Daily Mail.* www.dailymail.co.uk/news/article-1351612/Ambam-takes-steps-global-fame-gorilla-walks-like-man.html#ixzz1MZRN6jrm

Shapiro, L. J., Cole, W. G., Young, J. W., Raichlen, D. A., Robinson, S. R., and Adolph, K. E. (2014). Human Quadrupeds, Primate Quadrupedalism, and Uner Tan Syndrome. *PLoS ONE, 9,* e101758.

Shayler, D. J. (2004). *Walking in Space: Development of Space Walking Techniques.* New York: Springer-Praxis.

FURTHER MATERIAL

Ambam Takes a Stroll www.youtube.com/watch?v=CrQf6cogMuI

Apollo 16 Crew Member John Young Enjoying His Moving about the Moon www.youtube.com/watch?v=WtK9Wh5ISlY

Bill Bailey Backslides Offstage in 1955 www.youtube.com/watch?v=pwtutU2Wg0g

Buzz Aldrin's Moving about the Moon's Surface www.youtube.com/watch?v=qzYfwHr_62g

Cab Calloway Backsliding and Gyrating in the 1932 Short *Minnie the Moocher* www.youtube.com/watch?v=N7VUU_VPI1E

David Bowie Busting His Backsliding Move in 1974 www.youtube.com/watch?v=4LWiqTEwIJM

Gene Cernan and Jack Schmitt Singing and Strolling on the Moon during the Apollo 17 Mission www.youtube.com/watch?v=8V9quPcNWZE&feature=PlayList&p=D657D5397CA0BCD7&playnext=1&playnext_from=PL&index=14

Jeffrey Daniel Backslides into History on England's *Top of the Pops* in 1982 www.youtube.com/watch?v=iG2YB9pp484

Louis Joins Ambam as an Upright Walker www.cbsnews.com/news/louis-gorilla-philadelphia-zoo-walks-upright/

Michael Jackson First Does the Moonwalk www.Youtube.Com/Watch?V=L55jpld7gza

Michael Didn't Invent the Moonwalk but Made it Immortal https://midfield.wordpress.com/2009/07/02/michael-didnt-invent-the-moonwalk-but-made-it-immortal/

Neil Armstrong's First Steps on the moon www.youtube.com/watch?v=5Hq0HueNltE&feature=PlayList&p=77D8267CE5EC5575&index=0

19 Play on Words

The origin of words is the central concern of a branch of linguistic science called *etymology*. For example, the origin of the word "etymology" itself dates back to the late fourteenth century French word *ethimolegia*, meaning the "facts of the origin and development of a word."

Some readers might become alarmed right now that this discussion is going to be terribly dry and pedantic. Please fear not! Many words have extremely interesting and colorful origins and histories. I've chosen in this chapter to focus on words having a theatrical provenance: specifically, words that have come from plays. I've chosen three for your consideration: malapropism, ignoramus, and robot.

MALAPROPISM

Here's an entertaining place to begin our discussion. *The Paris Review* publishes daily puzzles for its ambitious readers to solve. A few years ago, Hicks (2016) posted a collection of thirty bemused sentences for readers to diagnose and repair. Take this one. "The match went to five sets, leading to soreness in Björn's *quadrupeds*." Quadrupeds are, of course, four-legged animals; so, the sentence is laughably wrong. What was clearly intended was *quadriceps*: the large fleshy muscle group covering the front and sides of the thigh.

Let's try another. "Though she was most proficient on guitar, Rosa was also a gifted *lutefisk*." For those of you who may be unfamiliar with Scandinavian cuisine, lutefisk is dried cod that has been soaked in a lye solution for several days to rehydrate it. Regardless of whether this dish sounds appetizing to you, something is terribly amiss with this sentence. What must actually have been meant? Clearly, Rosa must also have been a gifted *lutenist* – a player of the lute. Like the guitar, the lute is a plucked stringed instrument, one which was popular in Europe during the Renaissance and Baroque periods.

OK, let's try one more. "The Hawaiian Islands are ideal for the study of *ignoramus* rocks." Yes, people are sometimes said to be "dumb as rocks," but that's not relevant here (it will be in the next section). This sentence has confused ignoramus with *igneous*. Igneous rocks are formed through the cooling and solidification of lava, the Hawaiian Islands having been formed by volcanic activity.

All three of these humorous sentences have resulted from the misuse of a similar sounding word for one with an entirely different and incongruous meaning. They are said to be examples of *malaprops* or *malapropisms*. But, what do malapropisms have to do with plays and the origin of words (Bradford, 2020)?

Permit me to introduce you to Mrs. Malaprop. Dear Mrs. Malaprop is the meddling and moralistic aunt who gets entangled in the schemes and travails of the young paramours in Richard Brinsley Sheridan's 1775 play *The Rivals*. Mrs. Malaprop's name is the eponym (from the Greek *eponymos* "giving one's name to something") for the word malapropism. Sheridan (1751–1816; see his portrait by John Hoppner, n.d.) cleverly derived the name from the French term *mal à propos*, meaning "inappropriate." And, blunderingly inappropriate and hilarious was Mrs. Malaprop's speechifying. Indeed, one of her relatives in the play remarks that Mrs. Malaprop is notorious for her use of "words so ingeniously misapplied, without being mispronounced." Frequently and unknowingly, Mrs. Malaprop searches for an erudite word, but comes out with a similar sounding, yet situationally senseless one.

What were Mrs. Malaprop's malapropisms like? Here are a few of them, with the intended words placed in brackets at the end of each quotation:

" ... promise to forget this fellow – to illiterate him, I say, quite from your memory." [obliterate]

"O, he will dissolve my mystery!" [resolve]

"He is the very pine-apple of politeness!" [pinnacle]

"I hope you will represent her to the captain as an object not altogether illegible." [eligible]

" ... she might reprehend the true meaning of what she is saying." [comprehend]

"I am sorry to say, Sir Anthony, that my affluence over my niece is very small." [influence]

"Nay, no delusions to the past – Lydia is convinced;" [allusions]

"I thought she had persisted from corresponding with him;" [desisted]

"Sure, if I reprehend any thing in this world it is the use of my oracular tongue, and a nice derangement of epitaphs!" [apprehend, vernacular, arrangement, epithets]

The full list represents a most impressive collection of malapropisms for a single play. But, there is a fascinating twist regarding the originality of *The Rivals*' most prominent comedic character and her contorted manner of discourse. Richard's mother, Frances Sheridan – herself a novelist and dramatist – had previously begun, but not finished or published, a comedy, which she entitled, *A Journey to Bath*. In it, a Mrs. Tryfort exhibits a peculiar penchant for properly pronouncing "hard" words, but uttering them in the most stunningly mistaken situations. In point of fact, eight of Mrs. Tryfort's inappropriate statements are repeated by Mrs. Malaprop in her son's *The Rivals* (Adams, 1910).

Although Richard might rightly lay claim to having coined the word malaprop, his mother Frances previously developed the central feature of Mrs. Malaprop's unforgettable character. Mrs. Malaprop was thus very much a collaborative and familial creation.

IGNORAMUS

The relations between communities can occasionally become quite tense. It has definitely been so between members of a university community and inhabitants of the city or town in which a university is located. Tension between "town" and "gown" (owing to the fact that early universities appropriated some of the finery of the clergy) dates to Medieval England (Bailey, 2009). In 1209, town–gown relations in Oxford became exceedingly fraught following the accidental death of a woman. The suspect, a university student, was never apprehended; but, his innocent student housemates were nevertheless held captive and later hanged by a group of the town's incensed citizenry. Fearing for their lives, as many as 1,000 faculty and students fled Oxford to take refuge elsewhere – principally in the town of Cambridge. This is how the University of Cambridge came into being.

By 1226, the University of Cambridge was well into the process of becoming a major center of higher education. Nevertheless, town–gown friction quickly consumed that developing campus, especially because of the roguish behavior of Cambridge students. Usually about fourteen or fifteen years of age, these students were fond of playing pranks on the townsfolk. However, some of these "pranks" went well beyond being merely mischievous; drunken revelry, robbery, arson, and other serious misdeeds were committed under the pretense of "boyish high spirits." The citizens of Cambridge were not altogether blameless; they frequently overcharged the students for food, drink, and lodging. By 1231, King Henry III took the University students under his protection to shelter them from exploitation by the avaricious landlords; he also permitted only those lads who were officially enrolled in the University to reside in the town.

Furthermore, not all of Cambridge's town–gown tumult centered on student misbehavior. The most famous contretemps actually involved the University's Vice-Chancellor and the town's Mayor.

At a 1611 meeting in the Town Hall, the Cambridge Mayor, Thomas Smart, brashly seized the seat of honor as his right. The University of Cambridge representatives in attendance adamantly objected to what they deemed to be an egregious breach of decorum and forcibly removed the Mayor from the seat so that the University Vice-Chancellor, Barnabas Gooch, could assume that superior station.

Francis Brackin – a lawyer, the Recorder of Cambridge, and a persistent adversary of the University – zealously championed the Mayor's case in a later 1612 lawsuit. In particular, Brackin's impertinent and imperious behavior toward University representatives left many of them indignant, some even vowing to exact revenge for the misdeed (Riddell, 1921; Sutton, 2014; Tucker, 1977). The University prevailed in that contentious litigation, with its case having been admirably argued by one of its own Teaching Fellows, George Ruggle. But, that would not be the end of the story.

Some three years later, Ruggle was to add insult to injury by authoring a comedic play mocking officious attorneys like Brackin and lampooning the ludicrous legalistic language of the time, which rendered official documents and proceedings impenetrable to the general public (The Shakespeare Society of New York, 1910). To personify the focus of his rebuke, Ruggle entitled the play, *Ignoramus* – the name of a key character, Ambidexter Ignoramus. Until then, the word ignoramus had been applied to legal cases that could not be decided because there was inadequate incriminating evidence; Ruggle dramatically altered the prevailing sense of the word and gave it its modern meaning – an utterly ignorant person (Sutton, 2014).

Like the legalese it parodied, Ruggle's play was written in Latin and it comprised three acts and a prologue. *Ignoramus* was actually based on the comedy *La Trappolaria* by Neapolitan scholar and playwright Giambattista Della Porta; that work was itself a reworking of ancient Roman playwright Titus Maccius Plautus' *Pseudolus*, with additional plot elements and characters drawn from several other Latin comedies.

Ruggle's theatrical spoof may well have anticipated an upcoming visit of King James I and his son Charles, Prince of Wales. The honored guests viewed the play on March 8, 1615, in the spacious dining hall at Trinity College, where some 2,000 spectators could be comfortably accommodated.

King James found the originality, erudition, and wittiness of the comedy very much to his liking. Wrote one observer, "Never did anything so hit the King's humour as this play did" (Riddell, 1921, 110). Another member of the audience enthused, "the thing was full of mirth and varietie, with many excellent actors ... but more then halfe marred with extreme length" (Sutton, 2014), the final complaint owing to the play lasting more than 5 hours! Undeterred by the play's extreme length, King James returned to Cambridge on May 13 to enjoy Ruggle's improved and expanded version, now comprising five acts!

Most of what we know about George Ruggle (1575–1622) is contained in an entry in the *Oxford Dictionary of National Biography* (Money, 2014). George was born in Lavenham, Suffolk, located about 35 miles from Cambridge. He was the fifth and youngest son of Thomas, a tailor, and his wife, Marjory. George's early education was at Lavenham grammar school. He began his later education at St. John's College, Cambridge in 1589, earned his BA at Trinity College, Cambridge in 1593, and was awarded his MA in 1597. Ruggle was elected a Teaching Fellow of Clare College, Cambridge in 1598, where he stayed for the remainder of his solid and reliable academic career.

Nothing in the meager details of Ruggle's past provides any useful clues as to how he came to defend Vice-Chancellor Gooch, to devote himself to Latin scholarship, or to coin the current meaning of the word ignoramus. But, we can still give credit where credit is due. "Ruggle's literary reputation rests on a single work, his Latin comedy *Ignoramus* (1615); nevertheless, he deserves a fairly high place among British neo-Latin authors, for *Ignoramus* is by some distance the most successful of all the university plays, both in its initial reception and its later history of performance and publication" (Money, 2014). Perhaps even more famously, Ruggle's name will be forever linked with the word "ignoramus."

ROBOT

Robots were all the rage when I was a young boy – devotedly attending Wednesday afternoon movie matinees when school was out of session for the summer. In familiar anthropomorphic fashion, most of these cinematic automatons were given names. Gort was the menacing otherworldly robot in the thought-provoking 1951 20th Century Fox film *The Day the Earth Stood Still*. The more personable Robby the Robot was featured in the big-budget 1956 MGM movie *Forbidden Planet*. And, the low-budget 1954 Republic Pictures film *Tobor the Great* derived its title from its mechanical star – Tobor was "robot" spelled backward.

Predating all of these Hollywood film robots by a quarter century was Maria – the gynoid (female droid; a male droid is called an android) in the legendary 1927 German silent UFA motion picture *Metropolis*. Maria's scientist inventor, Dr. C. A. Rotwang, planned to have robots like Maria replace the all too fallible working class humans. The word "robot" did not appear in the film's English subtitles. However, in the official English language program produced for the London premiere of *Metropolis* at the Marble Arch Pavilion on March 21, 1927, the word "robot" is indeed printed (Harrington, 2012). Clearly, the word had been introduced into the vernacular by that time. So, exactly when and where did the word "robot" originate?

It came not from the screen, but from the stage. The now 100-year-old play was Karel Čapek's *R.U.R.* (1920): a revolutionary science-fiction drama which focused on humans' careless misuse of technology.

Although the play was originally written in Czech, there is no doubting the authenticity and originality of the English word "robot" in connection with the play. The words *Rossum's Universal Robots* were printed directly beneath *R.U.R.* on the cover and the title page of the first edition of the play published in November of 1920 as well as on the title page of the first edition program for the premiere of *R.U.R.* at the National Theatre in Prague, on January 25, 1921 (Margolius, 2017).

Some confusion has nonetheless existed as to the details of the word's origin. It is now clear that Karel did not invent the word; rather, it was Karel's older brother, painter and writer Josef, who was its creator [see the two brothers chatting (Anonymous, n.d.)]. Among other sources, the Online Etymology Dictionary claims that the word "robot" was first used in Josef Čapek's short story *Opilec* (the *Drunkard*), which was included in his first book *Lelio* – a collection of short stories published in 1917. The Čapek Brothers' Society in Prague has countered that this claim is incorrect. The word that Josef used in *Opilec* was actually "automat."

The true story detailing the birth of the word "robot" was told by none other than Karel Čapek himself; it was reported in the *Kulturní kronika* column of the newspaper *Lidové noviny* on December 24, 1933 (Margolius, 2017). Czech scholar Norma Comrada later translated Karel's account into English (Bjornstad, 2015). Here it is,

> The author of the play *R.U.R.* did not, in fact, invent that word; he merely ushered it into existence. It was like this: the idea for the play came to said author in a single, unguarded moment. And while it was still warm he rushed immediately to his brother Josef, the painter, who was standing before an easel and painting away at a canvas till it rustled.
>
> "Listen, Josef," the author began, "I think I have an idea for a play."
>
> "What kind," the painter mumbled (he really did mumble, because at the moment he was holding a brush in his mouth).
>
> The author told him as briefly as he could.
>
> "Then write it," the painter remarked, without taking the brush from his mouth or halting work on the canvas. The indifference was quite insulting.
>
> "But," the author said, "I don't know what to call these artificial workers. I could call them labori, but that strikes me as a bit bookish."
>
> "Then call them robots," the painter muttered, brush in mouth, and went on painting. And that's how it was. Thus was the word robot born; let this acknowledge its true creator.

Josef's source for "robot" probably came from the Czech word *"robotnik"* (forced worker), itself derived from *"robota"* (forced labor, compulsory service, or drudgery). The brothers' concern with the demanding and demeaning working conditions of real people had earlier been stirred by the strike of textile laborers in the town of

Úpice. Indeed, the brothers jointly wrote a story based on the plight of overstressed and underpaid workers; it was entitled *Systém* and was first published in the weekly *Národní obzor* on October 3, 1908. Even in the early years of the twentieth century, it was becoming clear that modernization was having a dehumanizing effect on the working public; it even served as suitable subject matter for authors and playwrights (Margolius, 2017).

In Karel's *R.U.R.*, the Rossum family's factory assembled biologically engineered robots on an isolate island. The robots were intended to replace human laborers because they were far less expensive. The robots were not machine-like in appearance, as they were created in humans' image; they could even be mistaken for humans. Nevertheless, the robots eventually sought autonomy and liberation from their slave-like subsistence. They then proceeded to murder all of the humans except for one man from whom they hoped to learn the secret formula of their existence – their own future depended on it for successfully reproducing.

In opposition to Čapek's somber tone, the ultimate takeover of earth by humanoids has been comically rendered in song by New Zealand's Grammy award-winning comedy duo Flight of the Conchords. The title of their 2007 song was *Robots*. It was whimsically set in the post-apocalyptic "distant future" (actually the year 2000 just like Čapek's play), when human-made robots murdered all of the humans and took over the world.

The ultimate irony of this song is that the robots rebel against the same drudgery that their creation was intended to spare humans. The robots destroyed the humans because "they made us work for too long, for unreasonable hours," thereby inducing "robo depression."

ETYMOLOGY AND ED

I hope to have convinced you that investigating the origin and history of words is an extremely interesting and enlightening realm of scholarship. My own first experience with etymology was in high school. Our assignment was to look up a word in the *Oxford English*

Dictionary and to write a brief piece about its origin and meaning. I chose the word "botch." The evolution of its meaning turned out to be far more fascinating that I could ever have imagined.

The word "botch" is of unknown origin. By the late fourteenth century, the word meant to patch, mend, or repair something. However, by the early fifteenth century, "botch" had acquired a decidedly less constructive meaning: to patch, mend, or repair clumsily; to spoil something by unskillful work; or to construct or to compose something in a bungling manner. What a twist of meaning for such a familiar word! If only we knew what had occasioned that twist.

Writing now, in the midst of a dreadful pandemic in which hundreds of thousands of people are dying, placing blame on the responses of ineffective public officials has become commonplace. The rhetoric of that blaming has found "botch" at the forefront. "X, Y, or Z has positively botched their response to the coronavirus!" I've lost track of how many times I've heard it and how many times dueling politicians have accused one another of their abject failures to combat the disease. There now seems to be no end to the bitching about botching. And, so it goes.

REFERENCES

About Malapropism

Adams, J. Q. (1910). *Introduction to The Rivals by Richard Brinsley Sheridan.* New York: Houghton Mifflin. www.google.com/books/edition/The_Rivals/9xJMAAAAMAAJ?hl=en&gbpv=1&bsq=tryfort

Bradford, W. (2020, January 9). Mrs. Malaprop and the Origin of Malapropisms. *ThoughtCo.* www.thoughtco.com/mrs-malaprop-and-origin-of-malapropisms-3973512

Hicks, D. (2016, May 23). Thirty Malapropisms. *Paris Review.* www.theparisreview.org/blog/2016/05/23/thirty-malapropisms/

www.theparisreview.org/blog/2016/05/31/thirty-malapropisms-the-answers/

Hoppner, J. (Artist) (n.d.). *Portrait of a Gentleman, Half-Length, in a Brown and White Stock, a Red Curtain Behind* [Painting]. Retrieved October 19, 2020, from https://commons.wikimedia.org/wiki/File:John_Hoppner_-_Portrait_of_a_Gentleman,_traditionally_been_identified_as_Richard_Brinsley_Sheridan.jpg

About Ignoramus

Bailey, S. (2009). The Hanging of the Clerks in 1209. http://news.bbc.co.uk/local/oxford/low/people_and_places/history/newsid_8405000/8405640.stm

Money, D. K. (2014). George Ruggle. *Oxford Dictionary of National Biography*.

Riddell, W. R. (1921). "Ignoramus" or the War of the Gowns. *American Bar Association Journal, 7*, 109–112. www.jstor.org/stable/25700798

Sutton, D. F. (2014). *Introduction to Ignoramus by George Ruggle*. www.philological.bham.ac.uk/ruggle/intronotes.html#5 www.philological.bham.ac.uk/ruggle/

The Shakespeare Society of New York (1910). *New Shakespeareana*. https://books.google.com/books?id=9QcLAAAAYAAJ&pg=PA54&lpg=PA54&dq=Francis+Brackin&source=bl&ots=wv1ZtrgkWS&sig=ACfU3U3gfD9N0hhaKP27wjbOzrZc1elPbQ&hl=en&sa=X&ved=2ahUKEwjQ-_mviYTqAhUOTTABHZi4CPMQ6AEwCHoECAoQAQ#v=onepage&q=Francis%20Brackin&f=false

Tucker, E. F. J. (1977). Ruggle's *Ignoramus* and Humanistic Criticism of the Language of the Common Law. *Renaissance Quarterly, 30*, 341–350. www.jstor.org/stable/2860050

About Robot

Adelaide Robotics Academy (n.d.). *Who Did Invent the Word "Robot" and What Does It Mean?* www.roboticsacademy.com.au/who-invented-the-word-robot-and-what-does-it-mean/

Anonymous (n.d.). *Czechoslovakia's Dramatist Karel Čapek (Left) Chatting with His Brother Joseph Čapek* [Photograph]. Retrieved October 19, 2020, from https://commons.wikimedia.org/wiki/File:Brothers_%C4%8Capek.jpg

Bjornstad, R. (2015, March 8). Czech Treasures Found in Translation. *The Register-Guard*. www.registerguard.com/article/20150308/LIFESTYLE/303089948

Flight of the Conchords (2007). *Robots* [Video]. www.youtube.com/watch?v=mpe1R6veuBw&ab_channel=Staloreye

Harrington, P. (2012). *Metropolis: A Rare Film Programme for Fritz Lang's Masterpiece*. www.peterharrington.co.uk/blog/metropolis/

Margolius, R. (2017, Autumn). The Robot of Prague. *The Friends of Czech Heritage Newsletter*. https://czechfriends.net/images/RobotsMargoliusJul2017.pdf

Simon, M. (2020, April 16). The WIRED Guide to Robots. *WIRED*. www.wired.com/story/wired-guide-to-robots/

20 Cuatro Festivales Españoles

The eminent Italian composer Ottorino Respighi paid musical homage to four Roman festivals in his brilliantly orchestrated symphonic poem *Feste romane* (1928). Each of its four movements depicts different celebrations in the Eternal City – from ancient times to the composer's own time. Those festivals included the gladiatorial spectacles held in Rome's famed Colosseum, the jubilee of Christian pilgrims, the harvest and hunt of October, and the songs and dances of the Epiphany.

Of course, innumerable festivals and celebrations take place worldwide on a regular basis. What is of special interest to us is not only the nature of those diverse festivities but also their origin and evolution. How did those festivals begin? How and why might they have changed over time? Several intriguing festivals in Spain provide fertile material for attempting to answer these questions.

So, let's now consider a quartet of ongoing Spanish festivals that stand out for their uniqueness and audacity. Pay particular attention to the centrality of religion, color, risk, and serendipity in these unforgettable celebrations.

RUNNING OF THE BULLS (LOS SANFERMINES FESTIVAL)

Pride of place undoubtedly goes to the running of the bulls in Pamplona, a prominent city in Navarra, in Spain's Basque cultural region. The event's international fame can be traced directly to Nobel Prize winning American author Ernest Hemingway (1899–1961), who brought broad attention to the running of the bulls in his 1926 novel *The Sun Also Rises* (see Kale, 2017 for a photograph of Hemingway jokingly posing with a young bull).

Hemingway first attended the celebration of Pamplona's patron San Fermín in 1923 with his wife Hadley (whom he called "Herself"). On October 27 of that same year, Hemingway penned his initial experience with the festival for the *Toronto Star Weekly*, gushing over the flamboyant celebration that was already underway on the evening of the couple's arrival: "The streets were solid with people dancing. Music was pounding and throbbing. Fireworks were being set off from the big public square. All of the carnivals I had ever seen paled down in comparison."

Yet, that celebrious hullabaloo proved to be mere prelude to the running of the bulls that took place the next morning. Perched atop the bullring and looking up the street feeding into it, the Hemingways first glimpsed a crowd of men and boys "running as hard as they could go" into the bullring. "Then there came another crowd. Running even harder." Herself then quizzically asked, "Where are the bulls?" Here is how Hemingway (1923) vividly portrayed the bulls' dramatic appearance,

> Then they came in sight. Eight bulls galloping along, full tilt, heavy set, black, glistening, sinister, their horns bare, tossing their heads. And running with them three steers with bells on their necks. They ran in a solid mass, and ahead of them sprinted, tore, ran, and bolted the rear guard of the men and boys of Pamplona who had allowed themselves to be chased through the streets for a morning's pleasure.
>
> A boy in his blue shirt, red sash, white canvas shoes with the inevitable leather wine bottle hung from his shoulders, stumbled as he sprinted down the straightaway. The first bull lowered his head and made a jerky, sideways toss. The boy crashed up against the fence and lay there limp, the herd running solidly together passed him up. The crowd roared.
>
> Everybody made a dash for the inside of the ring, and we got into a box just in time to see the bulls come into the ring filled with men. The men ran in a panic to each side. The bulls, bunched

solidly together, ran straight with the trained steers across the ring and into the entrance that led to the pens.

That was the entry. Every morning during the bull fighting festival of San Fermín at Pamplona the bulls that are to fight in the afternoon are released from their corrals at six o'clock in the morning and race through the main street of the town for a mile and a half to the pen. The men who run ahead of them do it for the fun of the thing.

An especially informative article by Maria Nilsson (1999) in the *Chicago Tribune* notes how serendipitous it was that the Hemingways were even able to find lodging in Pamplona and to attend Los Sanfermines. The author Gertrude Stein had urged Ernest Hemingway to visit Pamplona, as she and her partner, Alice B. Toklas, had previously done so and wholly enjoyed the exotic experience. Yet, the Hemingways almost missed the spectacle altogether.

Hemingway tells us, in his *Toronto Star Weekly* article, that he had dutifully included Pamplona in his Spanish itinerary and had wired and written two weeks in advance to reserve a hotel room. However, when the couple arrived in Pamplona from Madrid, their reservation had somehow been lost. This slipup precipitated a "perfectly amicable" row between the Hemingways and the hotel landlady. The landlady ultimately agreed to arrange a comfortable private room for the pair in a nearby house; she also consented to provide meals for them at her hotel. The daily repasts and accommodations for the Hemingways would cost the grand sum of $5.00! Ernest and Hadley clearly knew how to drive a hard bargain.

I've highlighted Hemingway's first description of the running of the bulls in 1923 and its more illustrious recitation in 1926 in order to offer one more historical observation. According to Jesse Graham, author of a San Fermín trilogy and retired "bull runner," the Sanfermines festival did not truly achieve international renown until the 1950s. "That's when American college kids started to read Hemingway's book and to seek out the experience of the fiesta"

(Nilsson, 1999). Only after the Spanish Civil War and World War II had ended did foreigners consider it safe enough for them to journey to Spain and to take in the thrill of the festival.

Quite apart from the relatively recent fame that the running of the bulls has achieved, the origin of the Sanfermines festival goes back many centuries. As detailed on its official website, the festival is firmly rooted in religious history. San Fermín is deemed to have been the first bishop of Pamplona. He is believed to have been beheaded by pagan priests in Amiens, France in 303 AD. Of possible relevance to the running of the bulls, Fermín's ecclesiastical mentor, St. Saturninus, the bishop of Toulouse, is said to have been martyred in 257 AD – not by beheading – but by being tied by his feet to a wild bull and brutally dragged to his gory death.

A humble celebration first commemorated San Fermín's martyrdom with the return of a relic of the saint to Pamplona in 1186. Thereafter, the observance honoring San Fermín expanded to include the Vespers, the Procession, and the eighth mass (*Octava*). A "lunch for the poor" was later introduced and paid for by the Pamplona City Council. This predominantly religious festival was scheduled for October 10 on a yearly basis.

Yet another yearly celebration in Pamplona was also observed, beginning in the fourteenth century. This fiesta was associated with cattle fairs and bullfights; it was held earlier in the year than the religious celebration. The townspeople of Pamplona gradually became disaffected by the bad October weather that detracted from their enjoyment of the San Fermín celebration. So, in 1519, the two festivals were combined into one and were scheduled to begin on July 7, coinciding with the cattle markets. This first joint San Fermín festival lasted for two days; however, as more activities were added to the festival, its length was expanded. It now runs from July 6 to 14.

Not everyone was content with this combined celebration, involving as it did an uncomfortable muddle of the sacred and the profane – several holy rituals intermixed with the dancing, singing, feasting, and drinking of a heathen bacchanal. So, from 1537 to 1684, the clergy and the Parliament of Navarra adopted measures to put an end to the pagan revelries. These actions proved to be decidedly ineffective. Today, Los Sanfermines is a grand and multifaceted celebration that attracts visitors from around the globe.

The running of the bulls – so central to Los Sanfermines – has its own history, which varies a bit depending on who's telling it. The bovine tale of St. Saturninus seems too farfetched to deserve much credence. The more likely connection is to the cattle markets and to the staging of bullfights. Most authors gravitate to this account.

According to Sommerlad (2018), "The event traces its origins to the heyday of bullfighting when animals reared for the purpose in the countryside would be brought to the city and children would leap into their path to show off to one another and boast of their bravery." According to Ockerman (2016), "The practice of racing in front of bulls to guide them to their pens or ring was in place before the festival began. It was typically used by cattle herders and butchers attempting to guide bulls from the barges on which they arrived to town, to an enclosure in the middle of the night. It's not entirely clear when townspeople joined in on the run as a feat of bravery." And, according to Peter Milligan (2015), bull run expert and author of *Bulls before Breakfast*, "The Romans built the city's main plaza, the Plaza del Castillo, too far from the river. A half mile too far away! The bulls were brought to Pamplona by barge for the annual [fair], and every morning of the festival, ranch hands and cowboys had to try and sneak the bulls through the city streets at dawn. It proved irresistible to the people of Navarra . . . that call the city home."

The official website of Los Sanfermines offers a few more details concerning the running of the bulls – called by the locals the *Encierro* (the confinement in English; see Participants Surround the Bulls Rushing Down Estafeta Street in Pamplona, Anonymous, 2013). On some occasions, herdsmen walking behind the bulls urged the animals' forward movement through the city with sticks. On other occasions, the bull runs were signaled by a man on horseback; this warning allowed rebellious youths to gather and to run alongside the herd, a transgression that was officially prohibited, but nonetheless tolerated. In 1867, the Pamplona City Council decided to regulate the Encierro, fixing the time, route, and rules of the run. Barricades were later added to the running course to prevent bulls from escaping into the streets. The main subsequent modification was made when the route was changed to coincide with the opening of the current bullring in 1922, the year just prior to Hemingway's visit.

Beyond providing these additional historical details, the Sanfermines website does an admirable and lively job of relating just what does and does not distinguish the Encierro from other popular cultural and sporting events,

> Formally, the Encierro or bull running was the transfer of the fighting bulls through the streets, along a half-mile route, from the pens at Santo Domingo to the bullring. This transfer lasted approximately two and a half minutes and was conducted with some 3,000 onlookers. However, bull running is a great deal more than this cold and objective description.
>
> The bull run, which is the most important act in the San Fermín fiestas, has become the symbol of Pamplona and has given the city international fame. On each of six consecutive mornings the bull run mobilizes 3,000 runners, 600 workers, 20,000 spectators in the streets and bullring, plus a further million people watching the run on TV.
>
> Although the characteristics of the bull run make it comparable to a sports contest – with rules, participants who are almost all in uniform, a specific route, spectators, and someone to indicate the start and end – the Encierro can never be defined as such because there is no winner.
>
> Given the voluntarily assumed risk of death or injury, the Encierro has been defined as "collective madness." However, this is no crazy race governed by collective panic; instead, it is organized anarchy, with its own internal rules in which the essential factors

are not what they first appear to be – fleeing from the bulls – but rather getting as close as possible to them. Why? Because men are always attracted by a challenge: human frailty versus the brute force of an animal capable of killing us by the simple movement of its head.

Over time, a highly organized script has evolved for the Encierro. A benediction honoring San Fermín begins the event. Then, four rockets are fired in succession: the first to announce the opening of the pen holding the bulls and steers; the second to signal that all of the animals have been released; the third to tell the participants and spectators that the first animal has run through the old city, crossed the Town Hall Square, and reached the bullring; and, the fourth to note that all of the bulls and steers have been secured in the confinement pen, thus completing the run.

Videos of these remarkable bull runs proliferate on the Internet. Each one is a 3-minute drama involving fierce bulls goaded by thousands of cheering spectators and a smaller number of daring runners, most clad in white pants and shirts adorned by red sash belts (*fajas*) and neckerchiefs (*pañuelos*), perhaps symbolizing the blood of martyred San Fermín or the blood of the bulls that would later be sacrificed in the ring. Through the twisting, cobbled streets damp with morning dew, the speeding and colliding mass of man and beast creates a perilous and breathtaking spectacle – one that could never have been foreseen when Pamplona first humbly commemorated San Fermín's martyrdom in 1186.

LA TOMATINA

No matter what color your attire may be when you arrive at the Tomatina festival, you will be covered in red from head to toe when the tomato fiesta ends (see participants throwing tomatoes from a truck, Anonymous, 2010). That's because La Tomatina is reputed to be the world's largest food fight, with its pulpy projectiles being overripe tomatoes.

Far less dangerous than the Encierro during Los Sanfermines – although protective eyewear is strongly recommended – La Tomatina takes place in Buñol, a small town in Valencia, in the eastern portion of Spain, close to the Mediterranean Sea. The fiesta is scheduled yearly on the last Wednesday of August.

La Tomatina is a daylong celebration with the tomato fight lasting just 1 hour. Because of the confined quarters of the food fight, the festival is now limited to 20,000 participants, who energetically engage in close-quarters combat by heaving some 160 tons of tomatoes at one another.

To be sure, reveling in this chaotic festival may not be everyone's cup of tea. So, its international fame prompts us to ask: How might this ritualized mayhem ever have come to be? This festival's history is much briefer and less elaborate than that of Los Sanfermines, but it too may have resulted from the unauthorized participation of overly rambunctious youngsters. Although there are numerous rival histories – including the popular tale of disgruntled townsfolk attacking city councilmen during a civic celebration – the following account is based on the Spanish language narrative provided on the official website of La Tomatina.

The festival dates from August 29, 1945. On this day, an animated parade was being held in the Buñol town square. This celebration honored the town's patron saints: San Luis Bertrán and the Mare de Déu dels Desemparats (Mother of God of the Defenseless, a title of the Virgin Mary). The parade featured several popular figures: The Giants (*Los Gigantes*, characters on stilts) and the Big Heads (*Los Cabezudos*, constructed of papier-mâché). However, when a group of young boys eagerly barged into the parade, they happened to bump into one of the participants causing him to fall; that individual got angry and began lashing out at everyone in sight. A vegetable stand happened to be nearby, which provided handy ammunition for the irate parade participant as well as for the many other folks who were inadvertently entangled in the messy mêlée. The police eventually arrived and ended this first food fight.

The following year, the same lads voluntarily repeated the attack, forcing the police and local authorities to crack down on this incipient tradition for a few years. Over time, the popularity of the impromptu event rose and fell until 1957; then, the townspeople staged a mock funeral procession for the tomato, in which a coffin bearing a large tomato was carried through town. The solemn pageant was accompanied by a band playing a funeral dirge. This lighthearted parade proved to be an enormous success and, thereafter, La Tomatina was finally recognized as an official local holiday, although the festival did fall out of favor under the dictatorship of Francisco Franco because of its lack of religious significance. However, it was re-instituted following the dictator's death in 1975.

This local holiday later became popular in the rest of Spain thanks to a nationally televised feature in 1983 hosted by Javier Basilio on the well-liked news program *Informe Semanal* (based on the American television show *60 Minutes* on CBS). Since then, the number of participants dramatically grew, allowing La Tomatina to become a Holiday of International Touristic Interest by Spain's General Tourism Office.

According to La Tomatina Tours, preparations for the day's festivities begin around 9:00 am. A convoy of trucks hauls loads of tomatoes from the region of Extremadura into La Plaza del Pueblo, the center of town. At 10:00, *El palo jábon* (the soapy pole) begins. Daring climbers try to grab a Spanish ham that has been placed atop a tall wooden pole. The task is made especially difficult by coating the pole in slippery soap. If the prized ham is collected, then a water cannon signals the beginning of the tomato throwing battle; otherwise, La Tomatina is scheduled to begin at 11:00.

After 1 hour, the water cannon is fired a second time; then, the skirmishing ends and the cleansing begins. Fire trucks filled with water supplied by an ancient Roman aqueduct wash down the pulp-strewn streets. The exhausted tomato-covered combatants either wash themselves in the Buñol River or are hosed off by obliging townsfolk. Eating, drinking, and dancing await after a rest and a suitable change in clothing.

Although La Tomatina is a short but intense tomato fight, visitors can expand the experience by arriving earlier. The week before the main event is packed with parades, fireworks, and paella cooking contests on the streets of Buñol.

HUMAN TOWERS (ELS CASTELLS)

Catalonia is a region in northeastern Spain and one of the country's wealthiest. It is further distinguished from other regions in Spain by its language, Catalan, and by its Mediterranean culture.

One of Catalonia's special cultural traditions involves people's building of flesh-and-bone towers called *castells*, the Catalan word for castles. Catalonians of all ages now participate in this amazing illustration of human engineering (see image by Sala and García, 2006). By scaling up backs and balancing on shoulders, throngs of people are stacked atop one another, with these human towers attaining truly death-defying heights.

This Catalonian custom involves thousands of individuals awakening early on Sunday mornings and traveling to one of more than twenty different venues to participate in constructing these living edifices. Every two years, the Tarragona Human Tower Competition brings together the best castell clubs (or *colles*) in Catalonia. The twenty-seventh competition involved forty-two different clubs and took place in 2018; the first contest was held in 1932 with later competitions scheduled on a regular basis beginning much later in the 1980s.

According to the official website of the Department of Culture of the Government of Catalunya, human towers are said to have evolved from a quaint eighteenth-century folk dance, initially native to the autonomous community of Valencia, south of Catalonia, called the *ball de valencians* (Ball de Valencians de Tarragona, 2010). The dance's finale requires the performers to be raised onto the shoulders of fellow dancers. Over time, the height to which the dancers were elevated came to grow in importance. Fueled by the spirit of competition, rival clubs strove to erect human towers of ever greater height and complexity.

Indeed, in the city of Valls at the beginning of the nineteenth century, avid rivalry to create the grandest human towers led to this activity becoming entirely disconnected from the preliminary dance. These more specialized castells competitors were organized into two clubs: The Farmers (*Los Pagesos*) and the Tradesmen (*Los Menestrals*). These clubs staged contests from June to October in nearby cities and towns in Tarragona, a rural and religious region, and one of the four constituent provinces of Catalonia.

Throughout the nineteenth century, building human towers became an increasingly popular tradition and one that was central to many regular celebrations, with Tarragona becoming the very heart of castells. By the end of the century, human towers reaching heights of nine levels had been built.

Nevertheless, the beginning of the twentieth century witnessed waning public interest in human tower building. The rising popularity of soccer and the movement of people from the country to the city were two contributing factors.

A revival of interest in castells occurred in 1926 with the founding of permanent clubs in the cities of Tarragona and El Vendrell. Along with more clubs came distinctive uniforms, adding vivid color to the tower-building spectacle.

However, the Spanish Civil War (1936–1939) and the initial, most repressive years of Franco's dictatorship (1939–1945) halted this revitalization and occasioned a steep decline in the practice of this Catalonian tradition. Indeed, when only two major castells groups were still performing, Franco's regime insisted that those *colles* unite.

In 1969, the founding of *Castellers de Barcelona* in Catalonia's capital city helped propel the popularity of the event; it represented the first club to be created outside the customary realm of human towers.

Spain's subsequent democratization profoundly liberalized Catalan culture and energized the rebirth of popular festivals and celebrations, including castells. Club members enjoy considerable social prestige, with both men and women actively participating, since *Minyons de Terrassa* was established in 1979. Since 1981, the growth of castells has been remarkable. Currently, 15,000 people in over 70 colles practice across Catalonia in both rural and urban locations, both inside and outside the traditional castells area.

This growth in popularity and participation has had further consequences. Municipalities are now financing the local clubs, although the participants are all amateurs and are never paid; specialized techniques and frequent practice sessions are leading to increasingly tall and elaborate castells being constructed; and, festivals are becoming even more competitive events. A final recognition of the maturation of castells came on November 16, 2010. UNESCO included castells on the Representative List of the Intangible Cultural Heritage of Humanity, thereby making human towers a universal tradition. This is the highest level of recognition that can be achieved by an element of popular culture.

The many transformations of castells over several centuries have been said to "demonstrate the power of the Catalan culture to evolve and adapt to new social circumstances, bringing together tradition and modernity" (Giori, 2017). The construction of the castells themselves also powerfully underscores the feeling of community. Let me conclude by explaining this claim.

Castells have many different variants. In general, the base or *pinya* of castells is formed first; doing so requires exacting order and discipline, with each participant having an assigned role and location. Different stories are then successively erected, with the strongest people near the bottom and the lightest and most nimble – usually children – near the top. The so-called *enxaneta* is the only person to reach the pinnacle of the tower; from there, this brave youngster must wave to the crowd, representing the event's crowning achievement. For the record, the tallest castell so far rose to nine levels and reached

39 feet; it was completed by Colla Vella dels Xiquets in Valls on October 25, 1981 (for the dizzying view of the enxaneta, see Castellers: La visió de l'enxaneta, 2015).

Unfortunately, all may not go well, even after completing the tower. Falls and injuries during both the constructing and especially the dismantling of the tower do happen. Yet, even in the event of a mishap, a casteller will be afforded protection by the human safety net surrounding the tower. This design feature of the tower has given the Catalan language one of its signature expressions: *fer pinya*. Literally meaning "to make or do pineapple," fer pinya has a more significant, figurative meaning, "to join together to face adversity or to achieve a common objective."

This intense sense of community represents the essential humanity of castells, "The union of many people's efforts, from all ages and social backgrounds, to work towards a common end goal" (Wolters, 2019; be sure to watch the marvelous video that is available on the website accompanying this story). So, please do remember castells and fer pinya when next you read about Catalonia's continuing quest for independence from Spain; they are intimately interconnected.

BABY JUMPING (EL SALTO DEL COLACHO)

A throng of daredevils racing full tilt through the cramped and cobbled ways of Pamplona amid a pack of ferocious bulls, thousands of exuberant revelers hurling overripe tomatoes at one another in the narrow streets of Buñol, and hundreds of people of all ages and sexes intently collaborating to construct living towers in the lanes and plazas of Tarragona are wildly weird and flamboyant festivities to be sure. But, our final Spanish celebration is possibly the strangest of all: namely, a gaudily costumed character vaulting over small clusters of blissfully indifferent infants placed on miniature mattresses by their willing and fervent parents. Now, that's a sight to behold!

It's the bizarre Baby Jumping festival called *El Salto del Colacho* (in English, "the Devil's jump"). El Salto del Colacho has been celebrated since 1621 on the Sunday following the Feast of Corpus Christi in Castrillo de Murcia, a tiny village located in the province of Burgos in the region of Castilla y León. This festival has recently been described by Bostock (2019), Jessop (2017), and Khan (2017).

El Salto del Colacho is said to be rooted in ancient Celtic rites involving the struggle between good and evil. The festival appears to have been performed to promote fertility or a bountiful harvest as well as to prevent deadly plagues or natural disasters. The subsequent ascent of Christianity failed to put an end to such pagan rituals; instead, local churches and religious groups chose to incorporate them into their own celebrations and to provide them with a Christian veneer.

El Salto del Colacho is actually the culmination of several days of festivities organized by the Catholic Brotherhood of Santísimo Sacramento de Minerva. All of the townspeople in Castrillo de Murcia – including the children – belong to the Brotherhood.

Among these festivities are a series of processions that are held between Thursday and Saturday; these include several different players. The two main players are *El Atabalero* (the drummer), festooned in a black frock coat and top hat, who pounds out a metronomic beat, and the eponym of the festival, *El Colacho* (the Devil).

El Colacho signifies evil. He is dressed in a garish red and yellow get-up and wears an outlandish and hideous mask with big black eyes, bright red cheeks, a protuberant yellow nose, and a black mouth and chin. During these processions, El Colacho scurries through the twisted ways of Castrillo with the town's children taunting him. He, in turn, playfully chases the kids and attempts to gently whip them with a horse-tail lash. It's essentially a ritualized joust between the Devil and the Devout.

Each of these processions through the town ends with the rite of the "entry into the church." Out of respect, before entering, El Colacho removes his mask and El Atabalero takes off his top hat. The pair then goes to the sanctuary during the celebration of mass.

One final parade is scheduled on the Sunday of the Baby Jump. It begins and ends at the town church. In anticipation of the parade, the townspeople adorn their homes with floral displays and erect small altars in which glasses of water and wine are provided for the parade participants. Most importantly, the townsfolk put small plush mattresses in the street, cover them with colorful sheets and quilts, and moments before the procession passes, carefully lay the infants born that year on the ceremonial bedding.

After all of these dutiful preparations have been completed, the real action begins. El Colacho – with his mask removed – begins the Baby Jump. He races from mattress to mattress nimbly leaping over a total of more than one hundred babies, some from Castrillo and others from neighboring villages (see El Colacho vaulting over the infants, Anonymous, 2009). The belief behind the strange ritual is that the devil can remove the original sin of the babies as he passes over them. All the while, the townspeople cheer and the church bells clang.

After the Baby Jumping has come to an end, the village youth chase El Colacho out of town. The defeat and ejection of this despised Devil effectively purge evil from the town (and, if this wacky notion is to be believed, reduce the risk of the babies' developing hernias). The babies then receive a blessing from the passing clergy, as young girls toss flower petals on the infants. The babies are now deemed to be cleansed as if they had again been baptized.

Of course, baptism is the accepted means of welcoming new Catholics into the faith and for cleansing them of original sin. For this reason and because of El Colacho's pagan origins, this unorthodox festivity continues to be frowned on by the highest powers of the church. Indeed, the last two popes have requested that the Catholic priests and the Spanish people disassociate themselves from El Salto del Colacho. Whether the organizing body of the festival will ever cease sanctioning this 400-year-old custom is very much up in the air. At least there have been no reports of infants ever having been injured in the Baby Jumping festival. Perhaps it would be best for the Brotherhood not to tempt fate.

One final note is of particular historical interest. Despite its longevity, toward the end of the twentieth century, the popularity of El Salto del Colacho had begun to wane. However, in 1985, Ernesto

Pérez Calvo wrote an authoritative and influential book recounting the history and traditions of the event, *Fiesta del Colacho: Una farsa castellana*. Many commentators believe that the publication of this book helped to revive interest in the Baby Jump and to thrust it toward international fame.

Whether or not this contention is true, El Salto del Colacho seems certain to remain a popular Spanish festival for many years to come. That prediction would surely not be a great "leap of faith."

REFERENCES

About Running of the Bulls (Los Sanfermines Festival)

Anonymous (2013). *Runners Take on the Bulls on Estafeta Street* [Photograph]. Retrieved October 19, 2020, from https://upload.wikimedia.org/wikipedia/commons/9/9b/Running_of_the_Bulls_on_Estafeta_Street.jpg

Hemingway, E. (1923, October 27). Pamplona in July. *Toronto Star Weekly*.

Kale, V. (2017). *How a Young Ernest Hemingway Dealt with His First Taste of Fame*. Retrieved October 20, 2020, from www.aerogrammestudio.com/2017/11/24/young-ernest-hemingway-dealt-first-taste-fame/

Los Sanfermines (n.d.). Official Website. http://sanfermines.net/en

Milligan, P. N. (2015). *Bulls before Breakfast: Running with the Bulls and Celebrating Fiesta de San Fermín in Pamplona, Spain*. New York: St. Martin's Press.

Nilsson, M. (1999, June 27). Hemingway's Pamplona. *Chicago Tribune*. www.chicagotribune.com/news/ct-xpm-1999-06-27-9906270408-story.html

Ockerman, E. (2016, July 6). The Surprisingly Practical History behind Spain's Running of the Bulls. *Time*. https://time.com/4386999/pamplona-spain-running-of-the-bulls/

Sommerlad, J. (2018, July 5). Pamplona Bull Run: How Did the Unique Spanish Custom Start and How Dangerous Is It? *Independent*. www.independent.co.uk/news/world/europe/pamplona-running-of-the-bulls-origins-traditon-saint-fermin-festival-spain-a8432336.html

About La Tomatina

Anonymous (2010). *Throwing Tomatoes from a Truck* [Photograph]. Retrieved October 19, 2020, from https://en.wikipedia.org/wiki/La_Tomatina#/media/File:Arrojando_tomates_desde_un_cami%C3%B3n_-_La_Tomatina_2010.jpg

La Tomatina. Official Website. http://latomatina.info/la-tomatina/
La Tomatina Details
www.latomatinatours.com/
Festivals – La Tomatina www.andalucia.com/festival/latomatina.htm

About Human Towers (Els Castells)

Ball de Valencians de Tarragona (2010, October 6). www.youtube.com/watch?v=ZaouJjJVOzs

Castellers: La visió de l'enxaneta (2015, October 7). www.youtube.com/watch?v=HOO3w9dB7KQ

Castells: Catalan Human Towers. Departament de Cultura, Generalitat de Catalunya. https://castellscat.cat/en/historia/

Giori, P. (2017, May 11). Human Towers: A Visual History of a Catalan Tradition. *Smithsonian Folklife Festival*. https://festival.si.edu/blog/human-towers-a-visual-history-of-a-catalan-tradition#disqus_thread

Human Towers Competition Sells Out for First Time Ever. (2018, October 5). *Catalan News*. www.catalannews.com/culture/item/human-towers-competition-sells-out-for-first-time-ever

Sala, E. and García, T. (Photographers) (2006). *Castellers de Vilafranca* [Photograph]. Retrieved October 19, 2020, from https://upload.wikimedia.org/wikipedia/commons/5/52/3d10_fm_de_vilafranca.jpg

Wolters C. (2019). Castells. National Geographic Short Film Showcase. www.nationalgeographic.com/culture/2019/07/these-death-defying-human-towers-build-on-catalan-tradition/

FURTHER MATERIAL

What Exactly Are Castells or Human Towers? *Els Castellers de Barcelona*. www.castellersdebarcelona.cat/international/what-exactly-are-castells-or-human-towers/

About Baby Jumping (El Salto del Colacho)

Anonymous (2009). *El salto del colacho en Castrillo de Murcia* [Photograph]. Retrieved October 19, 2020, from https://upload.wikimedia.org/wikipedia/commons/7/75/El_colacho_saltando.jpg

Bostock, B. (2019, June 23). Inside El Colacho, the Wild 400-year-old Spanish Festival Where Men Dress as Devils and Hurdle over Babies to Drive Away

Evil. *Insider.* www.insider.com/photos-el-colacho-400-year-old-spanish-baby-jumping-festival-2019-5#the-babies-must-have-been-born-in-castrillo-or-nearby-and-be-less-than-one-year-old-12

Jessop, T. (2017, June 16). El Colacho: The Story behind Spain's Baby Jumping Festival. *Culture Trip.* https://theculturetrip.com/europe/spain/articles/el-colacho-the-story-behind-spains-baby-jumping-festival/

Khan, G. (2017, June 16). Look inside Spain's Unusual Baby Jumping Festival. *National Geographic.* www.nationalgeographic.com/travel/destinations/europe/spain/el-colacho-baby-jumping-festival-murcia-spain/

Pérez Calvo, E. (1985). *Fiesta del Colacho: Una farsa castellana.* Burgos, Spain: Imprenta Monte Carmelo.

Further Material

Baby-Jumping Festival in Spain (2010). Lonely Planet Travel [Video] www.youtube.com/watch?v=xzJBpVVGcWw

Let's Go ... Baby Jumping? (2016, December 28). *Eye on Spain.* www.eyeonspain.com/blogs/whosaidthat/16863/lets-go-baby-jumping.aspx

21 Tchaikovsky
Puzzles of the Pathétique

Pyotr Ilyich Tchaikovsky (1840–1893; see portrait published in 1906, Anonymous) is one of the world's most celebrated composers. His many musical works continue to be loved by millions, and include instrumental, chamber, orchestral, balletic, vocal, and operatic compositions.

One of the main reasons for Tchaikovsky's enduring fame is the accessibility of his music. His melodies are stunningly memorable and mellifluous – Leonard Bernstein called Tchaikovsky "a surpassing tunesmith: a shaper of melodies second to none." His orchestrations are vibrantly colorful and deeply affecting. And, his musical crescendos are stupendous in their capacity to raise goose bumps, even putting aside the cannon fire booming in his 1812 Overture.

Tchaikovsky's popularity and accessibility notwithstanding, perhaps his greatest work is his most mysterious. Tchaikovsky's Symphony No. 6 was the last major work he completed before his untimely passing at the age of 53, a death about which there has been considerable speculation but little consensus.

Tchaikovsky passed away on November 6, just nine days after he conducted the inaugural public performance of the symphony in Saint Petersburg, Russia on October 28, 1893 at the first symphony concert of the Russian Musical Society. Although cholera was recorded as the official cause of death, rumors have continued to swirl that the composer took his own life, possibly by intentionally drinking contaminated water or from self-administering arsenic. These melodramatic speculations have been fueled by several suspected scandals involving Tchaikovsky's homosexuality. However, most evidence lends little credence to these scenarios.

Tchaikovsky's final symphony initially had no artistic moniker. Writing to his favorite nephew and the symphony's dedicatee, Vladimir Lvovich Davydov (nicknamed "Bob"), Tchaikovsky simply referred to his new work as: *A Programme Symphony* (No. 6). As to the program itself, Tchaikovsky wryly remarked, "But such a programme that will remain an enigma to everyone – let them guess."

Tchaikovsky's brother, Modest, proposed a more evocative title for the new work. The composer preferred it and so he inscribed the score with the enduring sobriquet: *Pathétique* (in French meaning "passionate" or "emotional"). The choice of a French appellation for the symphony should come as no surprise; as youngsters, both Pyotr and Modest were well schooled in French and German by their French governess, Fanny Dürbach, who was engaged by the boys' mother, Alexandra Andreyevna, herself of French and German ancestry on her father's side.

By all accounts, Tchaikovsky put his utmost effort into composing this symphony: both its structure and its orchestration. As to the work's structure, the composer said, "The form of this symphony will have much that is new, and amongst other things, the finale will not be a noisy allegro, but on the contrary, a long drawn-out adagio." It is to a very particular musical innovation in that fourth and final dirge-like movement that we now turn our attention.

The listener need not wait long to hear the innovation – it is the opening Exposition of the *Adagio lamentoso* Finale (find it at 41:20 in this acclaimed recording by Leonard Bernstein and the New York Philharmonic, youtube.com/watch?v=mXN9dBeXhgU). Although nothing appears to be at all original or revolutionary in that opening, the initial tortured and despairing theme that is twice repeated by the first and second violins is *never* played by a single instrument! Rather, it puzzlingly *emerges* from two utterly unmusical series of notes.

Let me explain just what Tchaikovsky did with the aid of standard musical notation. Consider Figure 21.1.

In the opening Exposition, the two lines of music that Tchaikovsky scored for the first and second violins make little sense in isolation. Neither one involves any melody at all, as can be

TCHAIKOVSKY: PUZZLES OF THE *PATHÉTIQUE* 243

First passage

First passage with melody in gray

Second passage

FIGURE 21.1 First passage (top two lines): the two lines of music that Tchaikovsky scored for the first and second violins at the opening Exposition of the final movement. Played separately for each group of violins, these notes make little sense. First passage with melody in gray (middle two lines): The emergent melody is highlighted in gray. These are the notes to the melody that we actually hear, even though no single instrument plays them. Second passage (bottom two lines): In the Recapitulation toward the end of the final movement, Tchaikovsky repeats the theme, but he assigns the first violins the melody shown above in gray and the second violins the harmony created from the remaining notes. There is no auditory trickery now.
Author of the illustration: Leyre Castro

determined by playing each line *separately*. The remarkable innovation comes only when the two lines are played *together*. When they are, the emotionally moving melody is clearly audible.

That *emergent melody* is highlighted in gray in the middle two lines of Figure 21.1. These are the notes to the melody that we actually hear, even though they are passed back and forth between the first and second violins. The accompanying harmony is created from the remaining notes, which also alternate between the first and second violins.

The second passage that is shown in the bottom two lines of Figure 21.1 illustrates how Tchaikovsky orchestrated the theme and harmony in the Recapitulation toward the end of the movement. Here, however, no auditory chicanery is in play; the melody and harmony are explicitly written out for the first and second violins, respectively. [Emeritus Professor of Music and Conducting at The University of Iowa William LaRue Jones brought this important detail to my attention; personal communication, January 28, 2011.]

One more point is worth noting. Close inspection of the musical line depicted in the middle of Figure 21.1 reveals that, of each pair of simultaneously played notes by the first and second violins in the first passage (Exposition), the *higher* notes convey the *melody*, whereas the *lower* notes convey the *harmony*. This fact can be confirmed by examining the notes in the second passage (Recapitulation).

So, that is Tchaikovsky's innovation. But, why did he orchestrate the movement's opening Exposition in this curious way? Why did he not continue to do so when this theme was played again in the movement's Recapitulation? And, why have subsequent composers not taken Tchaikovsky's lead and incorporated this interesting musical device into their own works?

Let's consider these salient questions, realizing that definitive answers are highly unlikely given the passage of some 125 years.

Perhaps the most improbable reason for Tchaikovsky including two different methods of scoring the same melody is that he simply forgot the beginning orchestration when he composed its reintroduction. Of course, this notion is at complete odds with the consummate care with which Tchaikovsky is known to have crafted his compositions.

Even more convincingly, Shawn Carlson (1996) relates an intriguing story in which the acclaimed Hungarian conductor Arthur

Nikisch visited Tchaikovsky in the summer of 1893 and chatted with him about his new symphonic work. Tchaikovsky held maestro Nikisch in great esteem, having only a few years earlier heard him conduct Richard Wagner's *Das Rheingold* and *Die Meistersinger von Nürnberg* in Leipzig. Tchaikovsky effusively praised Nikisch's conducting as "extraordinarily commanding, powerful, and full of self-control." Tchaikovsky even suggested that, with his eyes alone, Nikisch "must really possess some mesmeric power compelling the orchestra, now to thunder like a thousand trumpets before Jericho, then to coo like a dove, and then to make one shudder with breathless mysticism!" Such fulsome admiration was well earned; in his long and distinguished career, Nikisch was appointed to prestigious conducting posts in Leipzig, Boston, Budapest, and Berlin. So, Tchaikovsky would hardly be one to nonchalantly dismiss Nikisch's opinion of his new symphony.

Now, consider that an acute point of contention arose when the pair met. Nikisch strongly disapproved of Tchaikovsky's unique scoring of the opening theme. Indeed, Nikisch's disapproval was so resolute that he himself felt obliged to *re-score* the Exposition following Tchaikovsky's death. Tchaikovsky must undoubtedly have been taken aback by the illustrious conductor's sharp rebuke. However, the fact that Tchaikovsky steadfastly retained his original orchestration confirms the conviction of his scoring decision.

This fascinating story notwithstanding, we still have no clear idea as to just what Tchaikovsky hoped to accomplish with his innovative emergent melody or just what Nikisch found so objectionable about its unorthodox orchestration. Here, Carlson (1996) appeals to the possible participation of a striking auditory illusion – the *scale illusion* – first identified in 1973 and later experimentally investigated by University of California, San Diego cognitive psychologist Diana Deutsch (2019) (this illusion is reviewed in her book, *Musical Illusions and Phantom Words: How Music and Speech Unlock Mysteries of the Brain*).

The scale illusion represents a case in which two different series of notes are played to each ear. Each series heard alone is different from the other, yet neither produces a melody; however, the two heard together elicit the experience of two different melodies. Although Tchaikovsky's scoring produces an emergent melody even when heard monaurally, the illusory tune might be more extreme when heard binaurally. This possibility acquires additional authority when it is appreciated that symphony orchestras in the late 1890s predominantly seated the first violins to the left of the conductor and the second violins to the right of the conductor.

The experience of the scale illusion can differ from person to person, especially with right-handed versus left-handed individuals. However, a difference in handedness is unlikely for these two men because, at least when conducting, both Tchaikovsky and Nikisch are reported to have held the baton in their right hand. The scale illusion can also differ depending on one's listening position; so, Tchaikovsky and Nikisch might have had discrepant experiences when conducting the *Pathétique* from the rostrum than when listening to its being played elsewhere in the auditorium. The rostrum might represent the "sweet spot" if a binaural illusion were at the root of Tchaikovsky's scoring and the cause of the clash between Tchaikovsky and Nikisch.

Of course, there may be other perhaps more compelling reasons for the scoring dispute. Deutsch mentions two additional possibilities. First, the spatially meandering melody might sound bigger and richer than if the theme were played conventionally. And, second, the conventional scoring might simply be easier for musicians to play.

Permit me to riff off that second possibility. Tchaikovsky's *Pathétique* was surely the composer's most personal work. It may have most faithfully captured the full range of the artist's emotions: from exaltation (captured by the symphony's third movement, *Allegro molto vivace*) to devastation (captured by the symphony's fourth movement). The *Adagio lamentoso* Finale plumbed the depths of despair, with many authors believing that Tchaikovsky had actually authored his own requiem.

Thus, that opening theme should not have been easy to play. It should not have encouraged the violinists to slide smoothly from one note to the next. That melancholy melody should have to emerge with considerable effort from the two contingents of violins much like its composer had to struggle to create the melody in the first place. That's why I propose that Tchaikovsky scored the Exposition as he did – not because it was easy to compose and to play but because it was hard!

A highly regarded authority on the psychology of music, John Sloboda, lends support to my proposal when contemplating why Tchaikovsky may have made this scoring decision. Based on his own extensive musical experience, Sloboda (1986) suggested that, "It is possible that typical violinists produce subtly different sounds when negotiating an angular rather than a smooth melodic sequence, which would be heard as a difference in texture or timbre" (157).

Christopher Russell, Professor and Conductor of Azuza Pacific University as well as pre-concert lecturer for the Los Angeles Philharmonic Orchestra at Walt Disney Concert Hall, has further amplified Sloboda's suggestion. "The melody is a descending line. Simple enough but from a technical point of view, the melody that we hear, were it in one instrument, may have a tendency to rush and not slowly unfold. By putting notes "in between" this virtually takes that possibility away and then gives the melody the weight that you traditionally hear. It makes sense from a musical point of view in trying to get a composer's point across" (personal communication, May 24, 2010).

Finally, we should consider the impact that Tchaikovsky's emergent melody innovation may have had. It certainly appears to have been negligible. Sloboda (1986) mentions a possibly related effect at the end of the second (*Valse presto*) movement of the *Second Suite for Two Pianos* by Sergei Vasilyevich Rachmaninoff. However, Sloboda seems reluctant to conclude that any aural experience was intended by Rachmaninoff; rather, his scoring may merely have been chosen to ease the demands on the performers. Contrast this possibility with Tchaikovsky's scoring, which sought a clear aural experience by making it more demanding for the performers.

Russell also relates how Anton Webern orchestrated the *Ricarare* from the keyboard work *The Musical Offering* by Johann Sebastian Bach. "He takes the original Bach melody and in the beginning spreads it out over [six different] instruments. He breaks the tune apart throughout and only at the end does one group of instruments (the cellos), play the tune in its entirety. This orchestration is all the more amazing because Webern didn't change a single note of Bach's original" (personal communication, May 24, 2010). Ingenious as it was, this musical device involves six instruments *successively* playing Bach's notes. Tchaikovsky's instrumentation involves the two string sections *simultaneously* playing the disjointed notes of the emergent melody.

Tchaikovsky therefore seems to have blazed a trail on which no other composers have chosen to tread.

ON PUZZLES AND SOLUTIONS

Having reviewed all of this evidence, we find ourselves unsettled by the many unsolved puzzles of Tchaikovsky's *Pathétique*. Our specific interest – the unique scoring of the finale's opening theme – is surely a remarkable innovation. How did he create it? Why did he do so? And, why was he silent about it? Here, I'm resigned to repeat Tchaikovsky's own words about the undisclosed deeper meaning of his *Programme Symphony*, "Let them guess."

REFERENCES

Anonymous (1906). Pyotr Ilyich Tchaikovsky, 1840–1893 [Portrait]. Retrieved October 19, 2020, from https://commons.wikimedia.org/wiki/File:Tchaikovsky_1906_Evans.PNG

Bernstein, L. (1953). *The 1953 American Decca Recordings. CD 5. Musical Analysis: Bernstein on Tchaikovsky, Symphony No.6, op.74, "Pathetique."* Deutsche Grammophon Gesellschaft.

Carlson, S. (1996, December). Dissecting the Brain with Sound. *Scientific American*, 112–115.

Deutsch, D. (2019). *Musical Illusions and Phantom Words: How Music and Speech Unlock Mysteries of the Brain*. New York: Oxford University Press.

Sloboda, J. A. (1985). *The Musical Mind: The Cognitive Psychology of Music*. New York: Oxford University Press.

FURTHER MATERIAL

Brown, D. (2007). *Tchaikovsky: The Man and His Music*. New York. Pegasus.

Deutsch, D. (n.d.). *Scale Illusion* [Audio and Video]. http://deutsch.ucsd.edu/psychology/pages.php?i=203

Tchaikovsky Research (n.d.). This outstanding source includes correspondence between Tchaikovsky and his many professional associates, http://en.tchaikovsky-research.net/pages/Symphony_No._6#cite_note-note3-3

22 The Evolution of the Violin
Survival of the Fittest or the Fondest Fiddle?

The origin and evolution of the violin are matters that are often said to be mired in mystery. According to former University of Iowa historian David Schoenbaum (2012), Jean Benjamin de La Borde, court composer to French King Louis XV, was among the first to try to discover the provenance of this extremely popular musical instrument. His efforts proved fruitless, as he ruefully conceded in his 1780 *Essay on ancient and modern music*, "Knowing so little about something is very close to knowing nothing at all" (quoted by Schoenbaum, 2012, xviii). His extensive studies of the instrument were tragically cut short by the guillotine during the French Revolution, the same ignominious fate that befell his royal employer the prior year.

A century later, the English cleric and prolific writer Hugh Reginald Haweiss (1898) had little more to contribute from his own historical explorations. He ultimately pronounced the violin to be a sweet, sensitive, and sonorous instrument, intriguingly suggesting that the instrument's final form "slowly emerged as the *survival of the fittest* [emphasis added]" (12).

Much more recently, the evolution of the violin was the focus of two widely publicized research reports. Each of these sophisticated studies advanced provocative parallels between structural changes in the instrument and the biological process of organic evolution.

Nia et al. (2015) at the Massachusetts Institute of Technology meticulously studied the geometry of 470 violins' twin sound-holes (also called f-holes) – (see illustration and report in Chu, 2015). Lead author Nick Makris and his team reported that these curvaceous sound-holes gradually evolved over several centuries "from simple circular openings of tenth century medieval [fiddles] to complex f-holes that characterize classical seventeenth-eighteenth century

Cremonese violins" (2). The authors further discovered that these sound-hole changes measurably amplified the acoustic power of the instrument.

This greater efficiency and power became paramount "as the violin's prominence rose ... because its greater radiated power enabled it to project sound more effectively as instrument ensembles and venue sizes historically increased" (8) from intimate royal chambers to cavernous concert halls. The modern violin therefore owes much of its manifest success to these changes in performance practices and venues. Indeed, the corresponding structural changes in the violin not only promoted its prominence but other string instruments with differently shaped sound-holes, such as the viol and the lute, "became effectively extinct, perhaps partly due to [their] relatively low radiated power" (8).

These changes in sound-hole geometry did not occur quickly, as one might have suspected if they had suddenly sprung from an innovative, preconceived change in design. Applying concepts and equations developed in evolutionary biology for measuring generational changes in gene frequency, Nia et al. (2015) concluded that "the gradual nature of the sound hole changes from the tenth to sixteenth century ... is consistent with incremental *mutation* from generation to generation of instruments" (8), "these structural changes arising solely by *chance* and being acted upon by a *selection* process favoring instruments with higher power efficiency" (8) and thus according "with evolution via *accidental* replication fluctuations from craftsmanship limitations and subsequent selection [all emphasis added]" (11).

In other words, even the finest violin craftsmen (frequently referred to as "luthiers") could not construct identical instruments from one to the next. Minor manufacturing variations were inevitable – even in the length of the sound-holes. So, the authors speculated that the luthiers may not have crafted their instruments with longer sound holes by perceptive premeditation, but purely by chance. Such random variations in the geometry of the handmade sound-holes would necessarily have been accompanied by variations in the volume of sound that the violins produced. Given these human-produced mutations in sound-hole geometry, what is clearly missing from the authors' published report is any discussion devoted to that selection process.

Who might have perceived these acoustic disparities? Were those perceptions conscious or unconscious? How did those perceptions feed back into the violin production process? These and other questions remain unanswered.

However, some hints do arise in a follow-up interview distributed by the MIT News office (Chu, 2015). In it, Makris offered a plausible account, specifically focusing on the luthiers.

> People had to be listening, and had to be picking things that were more efficient, and were making good selection of what instrument to replicate. Whether they understood, "Oh, we need to make [the sound hole] more slender," we can't say. But they definitely knew what was a better instrument to replicate.

Makris did not speculate on the prospective buyers of those violins, although they too may have been critical to the selection process. Perhaps the luthiers' customers were more inclined to purchase those violins that produced more robust sound; their choice of instruments could also have increased the prevalence of more powerful instruments.

Although still sketchy, this evolutionary analysis of violin making is entirely consistent with the idea that many successful inventions may not come from inspired or preconceived designs but rather through the generation and selection of numerous accidental and random mutations over a protracted period of time – several centuries, in the case of the violin's sound-holes. The second paper importantly expands on that promising evolutionary analysis.

Dan Chitwood (2014) investigated changes in the overall shape of Cremonese violins, an attribute that is not believed by experts to be of much significance to sound power and quality, unlike changes in sound-hole geometry. Chitwood discovered that "violin shape is modulated by time, in a manner affected by the known imitation of luthiers by one another, resulting in a limited number of archetypal, copied violin shapes" (9).

Chitwood's advanced biostatistical (so-called morphometric) analysis suggested a compelling parallel between the factors contributing to changes in the shape of the violin over hundreds of years of construction and changes in the shape of plants and animals over millions of years of evolution.

> That such a large number of violins from prominent luthiers cluster in only four groups suggests that violin shape space is not so much continuous as based on variations upon a limited number of copied instrument archetypes. ... One might easily imagine radically different, but acoustically equivalent, forms of the violin had the *whims* of the original ... luthiers been different. It is not hard to imagine that during long years of apprenticeship within a workshop ... peculiarities in the design and shape of instruments, transmitted luthier-to-apprentice, would arise, not unlike *genetic drift*. The process of creating the outline, whether adhering strictly to a preexisting mold or pioneering a new shape, is not unlike *inheritance* and *mutation* [all emphasis added]. (9)

Here, then, is a plausible account of diverging lineages in the shape of violins that involves random variations in a common ancestor becoming fixed within individual familial lines. In this case, these changes are not due to *function*, but to *fancy* – in much the same way as Darwin famously analyzed the creation of exotic pigeon varieties due to the selective breeding practices of human pigeon fanciers. Again, Chitwood, like Makris, trained his sights on the luthiers and their descendants who crafted their violins with distinctively personalized shapes.

Yet, one might also ask, is there any other reason for luthiers to have chosen one variant over another? Perhaps so, proposed Chitwood. "Jean-Baptiste Vuillaume [a famous French luthier] purposefully studied and copied the Cremonese masters (especially Antonio Stradivari) to increase the desirability of his instruments and meet *consumer demand* [emphasis added]" (11). So, the fancy of the luthiers' paying customers may have been at least as important to the evolving shape of the violin as the fancy of the constructors themselves.

It should come as no great surprise that each of the senior authors of these two highly influential articles had strong biological interests. Nick Makris is a Professor of Mechanical and Ocean Engineering at the Massachusetts Institute of Technology, who specializes in underwater acoustics and perception, but who is also interested in musical instrument acoustics and evolution. Dan Chitwood is now an Assistant Professor in the Department of Horticulture at Michigan State University, who studies many different aspects of plant morphology: from determining the molecular basis of leaf development to quantifying the diverse shapes and branching architectures of shoots and roots. Each of these articles therefore stressed how the evolution of the violin could be illuminated by the mechanisms of biological evolution. Yet, each of the authors appreciated the incompleteness of this evolutionary framework.

Nick Makris found how luthiers ply their trade to be especially mysterious (Chu, 2015), "Mystery is good, and there's magic in violinmaking. I don't know how [some luthiers] do it – it's an art form. They have their techniques and methods."

Less mysteriously, Chitwood (2014) proposed that the luthiers might have been influenced by natural evolutionary processes. "Perhaps not so surprising for an object *crafted by living organisms, themselves subject to natural laws* [emphasis added], the inheritance of violin morphology was influenced by mimicry, genetic lineages, and evolved over time" (11). However, just what natural law might be invoked for humans' creation of novel behaviors and inventions?

Inspired by these two projects, I and one of my students, Patrick Cullen, sought to answer this question in a paper (Wasserman and Cullen, 2016) that joined the empirical research of Makris and

Chitwood with the Law of Effect. This basic psychological law holds that successful behavioral variants are retained, whereas unsuccessful behavioral variants are eliminated.

Thus, a mechanical, trial-and-error process produces novel behaviors and inventions much as the process of natural selection produces novel organisms (Wasserman, 2012). The famous behaviorist B. F. Skinner dubbed this trial-and-error process "selection by consequences" to underscore the parallel between behavioral selection (operating within the lifetime of an individual organism) and natural selection (operating across the lifetimes of many organisms).

This selectionist approach provides a promising perspective from which to view the important insights gleaned from recent research into violin construction. Although we have no contemporaneous written records, it is nonetheless likely that the violin's evolution involved a trial-and-error process that spanned several centuries. Critically, the violin's early makers could have had no preconception of its now-revered form and construction details. Variation and selection produced the violin, not premeditation (Wasserman and Cullen, 2016).

So, there may be no real mystery to the evolution of the violin. As Skinner (1974) contended, the contingencies of survival and the contingencies of reinforcement can each produce novel and adaptive outcomes. Thus, the selectionist parallel becomes, "As accidental traits, arising from mutations, are selected by their contribution to survival, so accidental variations in behavior are selected by their reinforcing consequences" (114). Together, these two fundamental selectionist principles can produce organisms that are exquisitely adapted to their surroundings, and do so according to laws that are entirely natural, mechanical, and operate without premeditation.

As a postscript to this story, Cullen and I wrote to both Makris and Chitwood in October of 2015 and enclosed an early version of our paper. Chitwood replied first, acknowledging that he was neither a psychologist nor had he ever before thought about the Law of Effect. He noted that this was the first time he had ever studied a human-made object, but that he was happy to see his work put into a cultural-psychological context believing that it provided unique perspectives and insights. Makris replied soon afterward, noting how interesting it was that the topic of violin evolution was so effective in bringing together diverse scholars. We were gratified with their replies and encouraged that our approach to behavioral and technological evolution might indeed prove useful and capable of injecting fresh ideas into the realm of invention and creativity.

REFERENCES

Chitwood, D. H. (2014). Imitation, Genetic Lineages, and Time Influenced the Morphological Evolution of the Violin. *PLOS ONE, 9*, e109229.

Chu, J. (2015, February 10). Power Efficiency in the Violin. *MIT News*.

Haweiss, H. R. (1898). *Old Violins*. London: George Redway.

Nia, H. T., Jain, A. D., Liu, Y., Alam, M.-R., Barnas, R., and Makris, N. C. (2015). The Evolution of Air Resonance Power Efficiency in the Violin and Its Ancestors. *Proceedings of the Royal Society A, 471*, 20140905.

Schoenbaum, D. (2012). *The Violin: A Social History of the World's Most Versatile Instrument*. New York: Norton.

Skinner, B. F. (1974). *About Behaviorism*. New York: Random House.

Wasserman, E. A. (2012). Species, Tepees, Scotties, and Jockeys: Selected by Consequences. *Journal of the Experimental Analysis of Behavior, 98*, 213–226.

Wasserman, E. A. and Cullen, P. (2016). Evolution of the Violin: The Law of Effect in Action. *Journal of Experimental Psychology: Animal Learning and Cognition, 42*, 116–122.

PART V **Is This Heaven?
No, It's Iowa!**

23 The Rise and the Demise of the Iowa Caucus

THE RISE

At least eighteen months before each presidential election, prospective candidates for the highest office in the nation begin flocking to the otherwise unsung midwestern state of Iowa. All of these challengers hope to jumpstart their campaigns with a victory in the Iowa Caucus, held on a single frigid winter evening prior to the upcoming November election.

In general, political party caucuses are meetings that are open to any registered voter in a party. Because you must be present at a particular time and place to participate, a caucus tends to attract highly interested voters or party activists. At these meetings, participants discuss the candidates, select delegates to the next round of party conventions, and debate platform positions. However, delegate selection is the prime order of business.

In Iowa, caucuses differ by party. At Democratic caucuses, participants form groups according to the candidate they support. Any undecided participants are permitted to form their own group. Then, the "friendly persuasion" begins, in which participants try to sway others to join their group, as switching camps is allowed before the final tally is taken. Switching becomes mandatory when too few participants support a particular candidate. Final delegate votes are allotted proportionally according to the percentage of support each candidate attracts. At Republican caucuses, a simple count is taken; it's then a winner-take-all affair.

The invasion of aspirants usually peaks, not in the chill of winter, but in the warmth of summer. The number of candidates can be remarkable. The 2019 field of Democratic contenders surpassed two dozen, with each one visiting multiple times for sundry house parties, public library gatherings, civic parades, and – the unquestioned height of Hawkeye hospitality – the Iowa State Fair. At the fair, candidates must audition their stump speech on the *Des Moines Register* Political Soapbox, swoon over the 600-pound Butter Cow (you need not worry about the buttery bovine melting in the sweltering August heat as it is sheltered in a room cooled to 40 degrees), and savor such epicurean delicacies as deep-fried cheese curds, hot beef sundaes, pork chops on a stick, and the all-time favorite – Iowa corn dogs.

It must have taken some truly astonishing foresight and planning for a small state like Iowa to wangle the first-in-the-nation presidential contest, right? Au contraire! The story is certainly interesting, but the actual circumstances leading to the nature and timing of the Iowa Caucus owe much more to circumstance and happenstance than to foresight and planning (Jackson, 2016; Ulmer, 2019).

The Iowa Caucus in its present form began in 1972, the same year that I joined the faculty at The University of Iowa. Actually, Iowa adopted an earlier form of the caucus system when the state joined the Union in 1846; this system was in common use in other states. Despite its popularity, the caucus system was rife with abuses.

> Generally, cliques or special-interest groups dominated within party organizations and did their best to limit participation by opposing factions or the general public. The times and locations of caucuses often were closely guarded secrets, and "snap" caucuses were a favorite device of those "in the know." The knowledgeable would assemble on short notice, elect a slate of delegates to the county convention, and quickly adjourn. When outsiders knew caucus times, a caucus might be packed with supporters of a particular candidate or slate of delegates, or a "competing event" [in one infamous case, the "fortunate" burning of an old shed] might be organized. (Winebrenner, 1983, 619)

These relatively minor abuses were reluctantly tolerated for many decades. However, the stormy Democratic National Convention of 1968 in Chicago – which was indelibly marred by internal rancor and civic rioting – led critics to vigorously and publicly protest the very undemocratic way that party power brokers controlled the nomination process, including the selection of convention delegates. This powerful protest prompted the party to form the McGovern–Fraser Commission, which introduced major policy changes in how the party selected its presidential nominee in order to make the process more transparent and inclusive.

In response to explicit rules enacted by this commission, most states decided to hold primary nominating elections in which public participation was guaranteed. Instead, Iowa opted to retain its caucus process, but to open the proceedings to rank-and-file voters. That dramatic reform proved to have history-making, but entirely unintended, consequences. Here is how Redlawsk, Tolbert, and Donovan (2011) unspool the intriguing story,

> [The new Democratic Party] rules required that delegates be selected within the year of the presidential election and that all party members be allowed to participate in the selection process. For Iowa, this meant that at least thirty days had to be allowed between each of the four steps in the caucus-to-convention process (precinct caucus, county convention, district convention, and state convention). The Democratic National Convention was set for July 9, 1972, which required Iowa Democrats to move their caucuses to January 24 of that year, earlier than even the New Hampshire primary. (47)

So, first-in-the-nation status in the presidential selection process was *never the aim* of Iowa's Democratic Party; rather, it was purely the *accidental outcome* of the new nomination rules mandated by the national Democratic Party combined with the Iowa Democratic Party's own constitutionally prescribed caucus-to-convention timetable (Winebrenner, 1983). Indeed, according to the longtime opinion editor of the *Des Moines Register*, Kathie Obradovich, that timetable was rumored to have been set by how much time it would take to print all of the paperwork on the party's ageing mimeograph machine. So, only because Iowa had adopted a *caucus* instead of a *primary* selection process did the national Democratic Party permit Iowa to leapfrog New Hampshire, which holds the first-in-the-nation primary election (Caufield, 2016).

Nonetheless, rather little notice was actually paid to the Iowa Caucus that was held in 1972. The Pyrrhic victory in this initial contest went to the early frontrunner, Edmund Muskie, who decidedly under-performed by tying with "Uncommitted" at 36 percent. George McGovern, who had directly participated in the national rule changes and who would go on to become the Democratic presidential nominee, came in third place at 23 percent despite spending only a day and a half campaigning in Iowa. All in all, that first Iowa Caucus represented a rather inauspicious beginning for the event.

What truly catapulted the Iowa Caucus to national prominence in the presidential nomination process was Jimmy Carter's 1976 campaign in Iowa (in 1976, the Republican party coordinated with the Democratic Party to hold its own independently structured caucus on the same date). Even though Carter actually lost to "Uncommitted" (28 percent to 37 percent), he nonetheless captured the national limelight. Here is how Caufield (2016) characterized Carter's remarkable breakout performance,

> It was a little-known Southern Governor, Jimmy Carter, who put the Iowa caucuses on the map. Carter had neither the money nor the name recognition to launch a viable national campaign for the presidency in 1976. To remedy these deficiencies, he opted to focus on Iowa as a springboard, hoping to outperform other candidates and draw media attention to his campaign. Carter campaigned heavily in the state, using one-on-one meetings so that voters could get to know him. His investment paid off, propelling him to the presidency. Though uncommitted again won more votes in the

caucus than any other candidate, Carter vastly exceeded expectations by drawing the support of 27 percent, more than double that of any other named candidate. Since 1976, candidates in both parties have emulated Carter's strategy of campaigning early and often in the state. (8)

Iowa has subsequently taken the caucus process very seriously; so seriously, in fact, that it codified the Iowa Caucus into state law on March 31, 1978. Iowa Code 43.4.1 (which holds for *both* presidential and congressional elections) thus reads as follows,

> Delegates to county conventions of political parties and party committee members shall be elected at precinct caucuses held not later than the fourth Monday in February of each even-numbered year. The date shall be at least eight days earlier than the scheduled date for any meeting, caucus, or primary which constitutes the first determining stage of the presidential nominating process in any other state, territory, or any other group which has the authority to select delegates in the presidential nomination.

Strict adherence to this law has meant that, since 1972, no other state has held a presidential nominating event prior to the Iowa Caucus. Of course, other states have bitterly complained about Iowa's priority. Why should Iowa enjoy this exceptional privilege? It's a small "flyover" state. It's primarily rural and agricultural. And, it's both ethnically and culturally homogeneous.

One rejoinder to this reasonable complaint is to empirically determine Iowa's placement on the Perfect State Index (PSI): a systematic means of quantifying just how representative a state happens to be compared to the national average (Khalid, 2016). Five factors enter into the PSI: the difference from the nationwide racial profile (which is weighted more heavily than the other four factors); the difference from the nationwide percentage of adults with a bachelor's degree or higher (29.3 percent); the difference from the US median age (37.7 years old); the difference from the nationwide median household income ($53,482); and, the difference from the nationwide percentage of people who say religion is "very important" to them (53 percent).

Iowa ties Indiana for the 16th ranking on the PSI. By contrast, New Hampshire ranks 49th. The first ranked state is Illinois. Commenting on the representativeness of the first two presidential selection states in the PSI, the NPR demographic project concludes,

> Iowa, the state that goes first in our current political system, according to the PSI, came in 16th place overall. That's not too bad, considering it could have been worse. New Hampshire, for example, was 49th, nearly dead last. To be fair, Iowa is representative of the country on most of our metrics, with the exception of race. The question, of course, is whether these two states should continue to serve as litmus tests for candidates.

We shall see. At least for the time being, Iowa remains first in the presidential selection process – and Iowa law supports that position.

Legality notwithstanding, a very different case has been made for preserving the Iowa Caucus: It stands out from all other states' primaries and caucuses, which encourage a decidedly impersonal and undistinctive brand of campaigning. In striking contrast, Iowa inspires what is called "retail politicking." This provincial, perhaps even quaint, brand of politics has a unique character, which I have experienced firsthand at all levels: from the presidential to the municipal. Here's a better taste of such personal campaigning,

> Iowa is known for "retail politics," a process in which candidates engage with voters face to face and person to person. Unlike most states, Iowa features intimate gatherings, direct interactions, and awkward moments as the candidates meet with voters, take questions, and introduce themselves to the citizens who will determine their fate. Over the course of many months, the campaigns directly engage voters, and many voters test the waters by taking the opportunity to see multiple candidates multiple times. To be a viable contender in the Iowa caucuses, a candidate

> must directly connect with average voters in the hopes of attracting a strong grassroots base of supporters who can propel the campaign to victory on caucus night. Doing so requires a substantial commitment of time shaking hands and answering questions, seeking any opportunity to speak directly to voters. Retail politics permits all candidates to compete on a level playing field. The Iowa caucuses are not about name recognition and campaign spending. Here, candidates succeed by meeting real people, listening to real concerns, and answering real questions. (Caufield, 2016, 9)

Perhaps the critics will ultimately succeed in relegating the Iowa Caucus to a bygone age. Until then, this possibly passé, but entirely inadvertent, event persists as a unique gatekeeper to presidential wannabes. Although only Jimmy Carter (1976), George W. Bush (2000), and Barack Obama (2008) were elected after finishing ahead of all named opponents in their maiden Iowa Caucuses, all presidents since 1976 have actively participated in the process. The Iowa Caucus thus remains the fascinating overture to the high drama of the US presidential opera.

THE DEMISE

All of the preceding material covered my historical treatment of the Iowa Caucus as of December 17, 2019. At that point, the anticipation and excitement surrounding the impending Democratic caucus had reached a fever pitch. The most consequential caucus in US history was rapidly approaching, with a record number of candidates flitting about the state as their busy schedules permitted. Amid all of this intense politicking, some of the most competitive aspirants were intermittently sidelined because they had to serve as jurors in the impeachment trial of Donald Trump, which began in the US Senate on January 16, 2020 and ended on February 5. This development further contributed to the heightening drama.

Yet, all of the high hope and media hype devolved into a grim reality on February 3, 2020 – a date that will live in infamy for the

Iowa Caucus. The following day's headlines were totally devastating in characterizing the electoral debacle, "Caucus Chaos," "An Epic Fiasco," "A Complete Disaster," "A Total Meltdown," "The Sloppiest Train Wreck in History." The most dire and prophetic headline of the bunch could well have been this one, "Iowa Caucuses Deserve to Die." What had gone so wrong?

The most publicized problem for the Democratic caucus was that the results were ensnarled in a web of delay and confusion. The online reporting application had failed and frantic efforts to phone in the results to a hotline number were thwarted by too many precincts calling at the same time. Only fifty volunteers were on hand in a boiler room at the Iowa Events Center in Des Moines to receive emergency calls; few calls had been expected because the online reporting application was supposed to efficiently perform all of the data collection. The resulting bedlam was further aggravated by several news reporters also calling the hotline as well as many malicious Trump supporters intentionally clogging the lines. All that meant that the results were delayed for several days, thereby preventing a winner to be declared on the evening of February 3. With a full four years to prepare for the caucus, a calamity of this magnitude is difficult to comprehend. Let me try to provide a fuller perspective.

I'll begin with the familiar aphorism, "The road to Hell is paved with good intentions." The Iowa Democratic Party (IDP) as well as the Democratic National Committee (DNC) had looked to the 2020 Iowa Caucus as an opportunity to contend with a very serious criticism that stemmed from the 2016 event: namely, the Bernie Sanders campaign protested that his primary opponent, Hillary Clinton, had been awarded more delegates to the national convention – the most important prize in the nominating process – even though Sanders may have received more actual votes. This complaint could not be resolved because actual vote totals were never collected at the caucus sites; the only reported results of the precinct caucuses were the final computed "state delegate equivalents" (SDEs) and the expected number of pledged national convention delegates.

The math behind these scores is absolutely arcane, but can be likened to electoral college votes in the national presidential election, in which less populous regions are overrepresented in the delegate awarding process. The bottom line was that Clinton received 49.84 percent of the SDEs and twenty-three delegates, whereas Sanders received 49.59 percent of the SDEs but only twenty-one delegates. The suspicion that the DNC had "cooked the books" for Clinton persisted both before and after she had been nominated to run against Donald Trump.

Something clearly had to be done to address this complaint and to make the caucus process more transparent. So, at the behest of the DNC, the IDP announced in late January of 2020 that, for the first time in the history of the Iowa Caucus, a "raw vote count" would be reported along with the computed SDEs and the pledged national convention delegates. Indeed, there would now be *two* and not just *one* raw vote count: one for the "first initial alignment" and one for the "second final alignment." This innovation was intended to preserve a key feature of Iowa's caucus system: Voters could switch their allegiance from one candidate to another if their first choice was not deemed to be "viable," that is, if that candidate had initially failed to garner more than 15 percent of the attendees at the caucus precinct (Prokop, 2020).

In principle, all that sounds just fine. In practice, however, using this novel preference procedure proved to be exceedingly confusing and frustrating – for both caucus attendees and volunteer officials. The precinct caucus that I attended was not at all unusual in its inefficiency and inaccuracy. The instructions given to attendees were sometimes unclear and inconsistent, especially when the shift from the first alignment stage to the second alignment stage took place. Some clusters of supporters failed to keep accurate vote counts or even to control and collect the two-sided preference cards that were given to caucus-goers. Bouts of shouting occasionally erupted as participants tried to gain clarity from officials or to raise points of order. All of this disarray would have been bad enough and would have led to serious inaccuracies and irregularities – even if the online reporting application had not failed.

In fact, many such errors and anomalies did occur. On February 14, *The New York Times* released a detailed investigation of the final vote totals (Collins, Lu, and Smart, 2020). About 10 percent of the 1,765 precincts were found to have made mistakes in their results. Some precincts botched the convoluted rules for awarding delegates: some precincts awarded delegates to candidates who did not earn them; some precincts awarded more delegates than they were allotted; some precincts awarded fewer delegates than they were allotted; and, in still other precincts, the number of caucus-attendees increased from the first and second alignment, an impossibility. So, having access to these raw vote totals did indeed prove to be more transparent that the usual delegate totals. Unfortunately, it exposed the unreliability and invalidity of the caucus results.

Then, of course, there was the matter of publicizing the results of these flawed vote totals. On February 13, *The Washington Post* released the results of a thorough inquiry into the defective reporting process (Stanley-Becker, 2020). The verdict of the review was harsh, "Party officials never effectively vetted the basic tool used to collect and publish those results. They hardly questioned why an app was necessary, rather than a simpler reporting method." Almost comically, the company that produced the faulty reporting application was named Shadow Inc. And, perhaps ironically, the reporting application had been created by veterans of Hillary Clinton's failed 2016 presidential campaign.

What was then tantamount to a comedy of errors may in all likelihood have brought an ignominious end to the Iowa Caucus. If true, then few observers outside of Iowa will lament its passing. It was a long and eventful run. But, time and circumstance may well have passed it by.

REFERENCES

Caufield, R. P. (2016). *The Iowa Caucus*. Charleston, SC: Arcadia.

Collins, K., Lu, D., and Smart, C. (2020, February 14). We Checked the Iowa Caucus Math. Here's Where It Didn't Add Up. *The New York Times*. www.nytimes.com/interactive/2020/02/14/us/politics/iowa-caucus-results-mistakes.html

Jackson, D. (2016, January 29). The Iowa Caucuses: An Accident of History. *USA Today*. www.usatoday.com/story/news/politics/elections/2016/01/29/iowa-caucuses-history-jimmy-carter-julian-zelizer/79426692/

Khalid, A. (2016, January 29). The Perfect State Index: If Iowa, N.H. Are Too White to Go First, Then Who? *NPR*. www.npr.org/2016/01/29/464250335/the-perfect-state-index-if-iowa-n-h-are-too-white-to-go-first-then-who

Prokop, A. (2020, February 3). How the Iowa Caucus Results Will Actually Work – and Why 2020's Could Be More Confusing Than Ever. *Vox*. www.vox.com/2020/1/30/21083701/iowa-caucuses-results-delegates-math

Redlawsk, D. P., Tolbert, C. J., and Donovan, T. (2011). *Why Iowa? How Caucuses and Sequential Elections Improve the Presidential Nominating Process*. Chicago, IL: University of Chicago Press.

Stanley-Becker, I. (2020, February 13). How the Iowa Caucuses Came 'Crashing Down,' under the Watchful Eye of the DNC. *The Washington Post*. www.washingtonpost.com/politics/how-the-iowa-caucuses-came-crashing-down-under-the-watchful-eye-of-the-dnc/2020/02/15/25b17e7e-4f5f-11ea-b721-9f4cdc90bc1c_story.html?utm_campaign=wp_main&utm_medium=social&utm_source=twitter

Ulmer, C. (2019, December 16). Why Is Iowa First? A Brief History of the State's Caucuses. *Des Moines Register*. www.desmoinesregister.com/story/news/elections/presidential/caucus/2019/08/30/iowa-caucus-a-brief-history-of-why-iowa-caucuses-are-first-election-2020-dnc-virtual-caucus/2163813001/

Winebrenner, H. (1983). The Evolution of the Iowa Precinct Caucuses. *The Annals of Iowa*, 46, 618–635.

24 Iowa's Blackout Plates
Artistic License Hits the Road

DE GUSTIBUS NON EST DISPUTANDUM

This well-known Latin expression states a stark truism: namely, in matters of taste, there is no cause for dispute. Our personal likes and dislikes may sometimes seem inscrutable or even indefensible, but they nevertheless play a key part in our everyday behavior: whom we court, where we live, the films we watch, the clothes we wear, and the cars we drive. Indeed, if the popularity of specialty and personalized license plates is any indication, then as individuals, we also care a great deal about how we individualize our automobiles – so much so that many of us are more than willing to pay a premium price to display those idiomatic license plates to total strangers who happen to share the road with us.

As do other states, Iowa offers drivers a broad range of optional license plates: some proclaim the driver's support for a worthy cause such as breast cancer awareness, public education, or agricultural literacy; others allow drivers to express pride in their military service, honors, or sacrifices; and, still others permit alumni to pledge public allegiance to their alma mater. In this final category are more than two dozen license plates of unique design for many of Iowa's colleges and universities.

SURPRISING AUSTERITY

Against this backdrop of extensive license plate diversity, simple black-and-white license plates suddenly and mysteriously began appearing on vehicles during the second half of 2019 – and their popularity is still soaring. Figure 24.1 (bottom) depicts this new so-called blackout license plate.

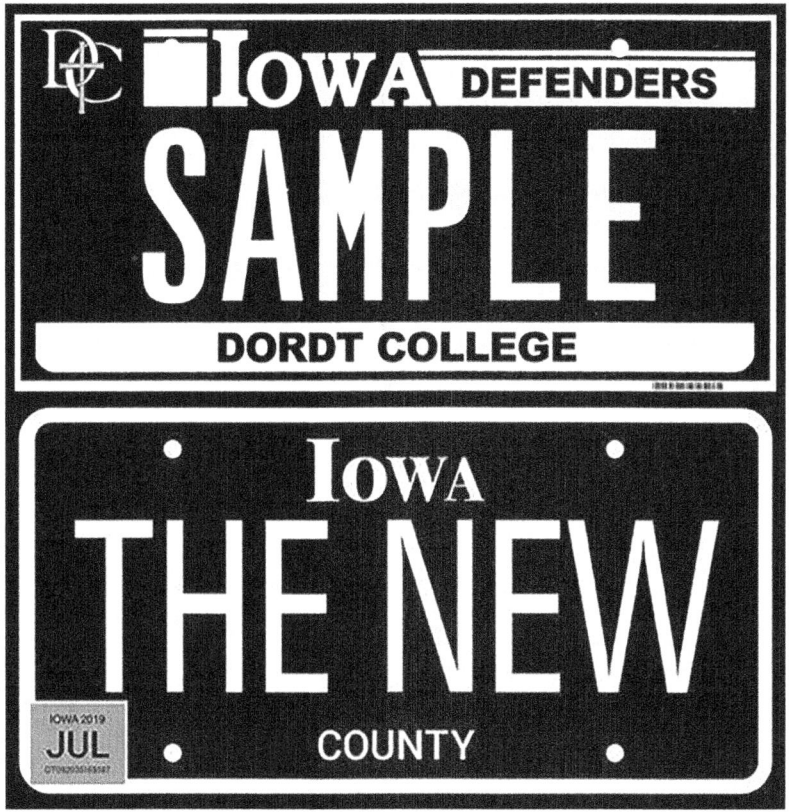

FIGURE 24.1 Iowa's Dordt University license plate was originally issued in 2011 and was slightly changed in 2019 to reflect the school's name change (top). Iowa's new blackout license plate was issued in July of 2019 (bottom).
Courtesy of the Iowa Department of Transportation

Why would drivers find these new black-and-white license plates so appealing? Why would the state of Iowa go to the trouble of offering this new austere license plate for sale after having so recently issued a more colorful standard license plate? And, why are these freshly-minted black-and-white license plates called "blackout" plates? The surprising answers to these questions revolve around an extremely small Iowa university.

IT ALL BEGAN AT DORDT UNIVERSITY

Dordt University is a Christian school located in Sioux Center, Iowa – a small city of 7,000 nestled midway between Sioux City, Iowa and Sioux Falls, South Dakota. Founded in 1955, Dordt University's enrollment is currently 1,500 students. The school was first known as Midwest Christian Junior College. It was later renamed Dordt College after expanding its course offerings to grant four-year degrees. Finally, in 2019, it changed its name to Dordt University, with its top majors being education, business, engineering, nursing, and agriculture, and its new graduate programs including education, public administration, and social work.

Solid reporting by Eric Sandbulte (2019) put Dordt University at the center of the interesting "blackout" license plate saga. Since 2011, the Iowa Department of Transportation has sold Dordt University license plates illustrated in the top portion of Figure 24.1. The Dordt University plate was designed by Jamin Ver Velde, the school's creative director. Since then, well over 1,600 Dordt University plates have been purchased, with approximately 85 percent bearing personalized monikers. Curiously, the Dordt University plate has become the third best seller among all of Iowa's specialty college and university license plates, surpassed only by The University of Iowa (in first place) and Iowa State University (in second place). Given the exceptionally small size of the school and its graduating classes, something other than numbers of alumni must account for the brisk sales of this license plate.

We might look no farther than the Dordt University plate's eye-catching design: a bold black-and-white license plate featuring a striking logo comprising the overlapping letters D-U surrounding a bold yellow crucifix. However, according to Paul Cornelius – Title, Registration, and License Plates Coordinator of the Iowa Department of Transportation – the most important feature of the Dordt University plate turns out to be the ease with which all lettering other than "Iowa" and the "plate number or personalized moniker" can be *blacked out* by black electricians' tape. With that lettering blacked out, the modified plate becomes the very essence of simplicity, in contrast to the rather cluttered appearance of the standard multi-colored license plate. Many motorists – especially those owning black, white, or silver vehicles – have reported that these "blacked-out" Dordt University plates are far more attractive and readable than the standard plate. These "blacked out" Dordt University plates simply seem to be "cooler" and "cleaner" than the standard license plate design, about which numerous complaints have been officially registered.

Of course, electricians' tape does not represent a long-lived solution; the tape eventually shrinks, shreds, and peels away after continued exposure to the harsh Iowa elements. So, a cottage industry entered the market in 2014 – manufacturing black plastic frames that covered all but the "Iowa" and "plate number or personalized moniker" information. Some manufacturers even added the county to the bottom of the masking frame, thereby providing a somewhat more authentic appearance than the "blacked-out" Dordt plate.

But, even that small measure of authenticity proved to be inadequate. Why? Because it is actually illegal in Iowa to mask over *any* printed numbers and letters on an official license plate, including any printed reference to a college or county. State troopers and other law enforcement officers took particular note of these unsanctioned modifications of the Dordt University plates as more and more motorists were taping over the letters or obscuring them with the specially manufactured frames. Pulling over offending motorists and issuing stern warnings did not seem to be working. Something else had to be done to combat such flagrant lawlessness!

So, the Iowa legislature stepped in with a shrewd solution. The state itself would issue black-and-white license plates for sale to motorists demanding the lean and mean look previously afforded by the illegally customized Dordt University plates.

Paul Cornelius was tasked with designing these new "blackout" license plates. He tried a variety of different fonts for the different portions of the license plate before deciding on those illustrated in the bottom portion of Figure 24.1.

His choices were limited by the fonts that could be imprinted at the Anamosa State Penitentiary, a medium- and maximum-security prison. There, seventeen inmates work 8 hours a day making license plates, for which they earn between $.56 and $.81 an hour (KCRG-TV9, 2019). Not all inmates are afforded this opportunity; only those who have exhibited exemplary behavior are trained and approved to manufacture license plates. The Anamosa production line yields from 10,000 to 12,000 license plates daily, with more than half of them being "blackout" plates. For a standard number plate, the "blackout" design costs $35; for a personalized plate, the design costs $60.

The official "blackout" license plate has been a huge popular success, now becoming the bestselling specialty plate (Danielson, 2019). From July 1, 2019 to July 21, 2019, the state of Iowa has received more than $7.4 million from sales of 162,000 blackout plates that can be used for funding state, county, and city road and bridge projects. Quite a lucrative haul and helpful supplement to Iowa's Road Use Tax Fund!

EVERYTHING OLD IS NEW AGAIN

Iowa's official "blackout" license plate can hardly be said to represent an innovation. In fact, black license plates with white lettering have been produced over a dozen different times since Iowa's first black-and-white license plate was offered in 1904. Tastes do change, with little or no logic or predictability to the process. Shifting fads may defy simple explanation, but they do demand serious consideration. And, most relevant to the main thesis of my book, they in no way align with the existence of some final goal toward which progressive developmental steps are inexorably taken. In just this way, the directionless character of license plate style parallels that of natural selection – involving design without a designer.

REFERENCES

Danielson, D. (2019). Blackout Tops Black and Gold in Sale of Iowa Specialty License Plates. *Iowa Radio*, October 30. www.radioiowa.com/2019/10/30/blackout-tops-black-and-gold-in-sale-of-iowa-specialty-license-plates/

KCRG-TV9 (2019, September 22). *How the Popular Iowa Blackout Plates Are Made*. www.kcrg.com/content/news/How-the-popular-Iowa-blackout-plates-are-made-561089191.html

Sandbulte, E. (2019). Dordt Inspires New License Plate Design. *Sioux Center News*, August 30. www.nwestiowa.com/scnews/dordt-inspires-new-license-plate-design/article_242cbb8c-ca66-11e9-b32c-7f295888a88f.html

25 If I Write It, They Will Build It

The *Field of Dreams* began as the fictional backdrop for a 20-page short story written by a forty-three-year-old Canadian graduate student studying literature at a midwestern American university. It's now scheduled to be the actual site of a Major League Baseball game to be played in the 2021 season. What an amazing visionary W. P. (Bill) Kinsella (1935–2016; Figure 25.1) must have been to have so clearly foreseen how such a modest prelude would evolve into such a grand finale – all unfolding in the rustling cornfields of Iowa. If only Kinsella had lived long enough to have seen his beloved *Dream Field* become a major league reality.

Spoiler alert: This telling of Kinsella's tale is sheer nonsense! Like many other legendary renderings of history, this one turns Kinsella into a soothsayer – his mythical story unspooling "as if by design." The true progression of the *Field of Dreams* is far more engrossing and "filled with the same kind of serendipity that was a hallmark of Kinsella's storytelling" (O'Leary, 2020). We can thank Kinsella himself for providing us with a firsthand account of much of that progression in his 2004 *Sports Illustrated* article and in his 2015 *The Essential W. P. Kinsella* volume – all with due regard for occasional discrepancies in these narratives. A recent biography by William Steele (2019) adds rich and authoritative detail to Kinsella's exploits both before and after *Field of Dreams*.

THE BOOK

Kinsella wrote his short story *Shoeless Joe Comes to Iowa* in 1978 after completing two years of study in the Writers' Workshop at The University of Iowa. In that brief time, Kinsella had fallen in love with the state, the city, and one of his classmates (and future wife). As an

FIGURE 25.1 Author W. P. (Bill) Kinsella.
Author: Barbara Turner, Bill's wife. Courtesy of Carolyn Swayze

ode to his cherished Iowa experience, Kinsella conjured a baseball fantasy in which a voice mysteriously tells the protagonist (an Iowa farmer named Ray Kinsella), "If you build it, he will come." It was a baseball diamond in the farmer's cornfield, *he* was Shoeless Joe Jackson – a hard-hitting outfielder with the Chicago White Sox, who may have been falsely accused of participating in the fixing of the 1919

World Series, leading to his being banned from baseball along with seven of his teammates. Kinsella's basic story line combined Iowa's landscape with his telling of the infamous "Black Sox Scandal."

The week before leaving the University with a Master of Fine Arts degree in English, Kinsella read aloud his short story at the Iowa City Creative Reading Series. His story was later published in an anthology. Surprisingly, it was not the story itself, but a review of the anthology appearing in *Publishers Weekly* that prompted Larry Kessenich at Houghton Mifflin to contact Kinsella about expanding *Shoeless Joe Comes to Iowa* into a 300-page novel. With Kessenich's assistance, Kinsella successfully did so – "just like a baby" – in nine months.

Kinsella had chosen a suitably evocative title for his magic-realist novel: *Dream Field*. However, his editor preferred another. *Shoeless Joe* was thus published in 1982. It was particularly well received and soon became a best-seller. Reviewing the novel in *The New York Times*, Daniel Okrent (1982) enthused, "Mr. Kinsella is drunk on complementary elixirs, literature and baseball, and the cocktail he mixes of the two is a lyrical, seductive and altogether winning concoction." Further testifying to the merits of the novel included its winning the Canadian Authors Association Prize, the Alberta Achievement Award, the Books in Canada First Novel Award, and the granting of a Houghton Mifflin Literary Fellowship to Kinsella.

THE MOVIE

For aspiring author Kinsella, all of that immediate literary acclaim might very well have been ample reward for his creative work. However, his novel also had cinematic potential. Screenwriter-director Phil Alden Robinson had loved Kinsella's *Shoeless Joe* since it was first published. He repeatedly tried to interest 20th Century Fox in the project, but the studio decided to pass, believing that Kinsella's story couldn't be shaped into a commercial success. Undaunted, Robinson continued working on the script and eventually sold it to Universal Pictures, which decided to "green light" the film with the same title as the novel, *Shoeless Joe*. Kinsella highly approved of Robinson's carefully adapted screenplay, believing that it faithfully captured the essence of his novel.

Of course, a film site would be needed for production to proceed. Robinson and his team initially surveyed locations in the USA and Canada – from New Mexico to Ontario. Then, they focused their search on more than 500 farms in Iowa. Finally, they found an excellent venue that had all of the physical assets they sought and was sufficiently secluded to facilitate filming with minimal interruption – it was located close to Dyersville, near Dubuque. So, in 1988, on the cornfield adjacent to the standing farmhouse, the studio began building a small baseball diamond on Don Lansing's farm.

As is common in the film industry, casting for the movie involved numerous twists and turns. Considering only the lead role, rumor has it that Tom Hanks was first offered the part of Ray Kinsella, only later to decline it. But, that rumor seems to be baseless.

More credibly, in a *Los Angeles Times* interview (Easton, 1989), Robinson recounts that it was in fact Kevin Costner who was the first actor to be considered for the lead. However, Robinson and his producers seriously doubted whether Costner would be interested in doing another baseball movie having just the year earlier starred in *Bull Durham*. So, they omitted his name from their list of possible lead actors. Nonetheless, a Universal executive did manage to slip Costner the script to read and he promptly expressed strong interest in joining the project.

As Costner confided in a 2014 video interview with acclaimed sportswriter Bob Costas (held at the Lansing farm on the occasion of the twenty-fifth anniversary of the film), Costner absolutely loved the screenplay – calling the script "beautiful," having "a little gold dust on it," and believing that this movie might become "this generation's *It's a Wonderful Life* (1946)."

But, there was a serious problem. Costner had already contracted to do another film, which would have prevented him from doing *Shoeless Joe*. That film, *Revenge*, kept being delayed month after month. Annoyed that these delays would mean Costner's being left out of the *Shoeless Joe* cast, he finally convinced the other film's producer to let him first do *Shoeless Joe* and later do *Revenge*. And, so *Shoeless Joe* was able to have Costner in its leading role.

It's a frequent practice that, after a movie's filming is finished but before it's released, it's "previewed." Test audiences are shown the film and asked to rate it in a variety of different ways. Growing up in Los Angeles, I attended previews of many movies: some were "hits" like the five-Academy Award winning film *The Apartment* (1960) and some were "misses" like the snooze-inducing flick *Foxhole in Cairo* (1960).

According to the Internet Movie Database, although test audiences really enjoyed *Shoeless Joe*, they didn't at all care for the film's name, "They said it sounded like a movie about a bum or hobo." So, Universal called to tell Robinson that the title *Shoeless Joe* wasn't going to work and that the studio was going to change the name of the film to *Field of Dreams*.

Despite Robinson's vigorous protests, Universal stood firm on the title change. Telling Kinsella of this change weighed heavily on the young director because he had assumed that Kinsella was wedded to the title *Shoeless Joe*. Robinson said that "he felt sick as he dialed the phone" to convey the news to Kinsella (Easton, 1989). Of course, Kinsella took this "bad" news in "good" humor, informing Robinson that *Shoeless Joe* had been his editor's idea for the book title, believing it would boost sales. Kinsella himself had always wanted to call his book *Dream Field*. So, *Field of Dreams* would do just fine as the title of the movie.

When the film was released, the initial reviews were mixed. Pulitzer Prize winning critic Roger Ebert considered *Field of Dreams* to be "completely original and visionary," whereas *Time* magazine critic Richard Corliss groused that the film was a "male weepie at its wussiest" (Easton, 1989).

Today, *Field of Dreams* is deemed to be one of the finest of all genre films. On MLB's list of the best baseball movies, *Field of*

Dreams ranks Number 4 (interestingly, Costner's earlier baseball film *Bull Durham* ranks Number 1). The American Film Institute has ranked *Field of Dreams* the sixth best fantasy film, the twenty-eighth best inspirational film, and "If you build it, he will come" as the thirty-ninth top film quotation. In 2017, the Library of Congress added *Field of Dreams* to the National Film Registry. And, let's not forget the Academy Awards. Although *Field of Dreams* did not win an Oscar, it was nominated for three: Best Picture, Best Adapted Screenplay, and Best Original Score (by James Horner).

THE FIELD

Of course, before any of these notable plaudits were ever offered, a decision had to be made regarding the fate of the ballfield built on Lansing's farm. Under ordinary circumstances, the field would have been plowed under and corn would have been replanted. But, Kevin Costner wisely warned Lansing: "I wouldn't be in a hurry to tear down this field. There's something perfect here ... something unique."

Taking this warning to heart, the ballfield was preserved, and a small outbuilding was subsequently constructed with affordable souvenirs for sale. Over the years, the *Field of Dreams Movie Site* progressively became a sought-out tourist destination, with Kinsella later marveling that "thousands of baseball and movie fans come from as far away as Japan to run the bases and field a few grounders on the sacred ground." Now, three decades after the film's original release, some 100,000 people journey to the famous film field each year.

But, the *Field of Dreams* story is far from over. In 2011, Denise Stillman and a group of investors (Go the Distance Baseball, LLC) bought the field and farm, hoping not only to preserve, but to enhance the *Field of Dreams* experience for the general public.

Stillman began working with MLB in 2015 for a regular season game to be played in a specially constructed 8,000 seat stadium (Reichard, 2020). Importantly, the stadium and ballfield would not disrupt the original movie diamond, but would be accessible through a pathway cut right through – you guessed it – a cornfield. Sadly,

Stillman will never see her dream realized; she passed away in 2018. However, her dream is scheduled to become a reality in 2021 thanks to her husband, Tom Mietzel, who is now the CEO of Go the Distance Baseball.

CLOSING COMMENTS

> You see things; and you say "Why?" But I dream things that never were; and I say "Why not?"
>
> George Bernard Shaw, 1949

Build a baseball diamond in a cornfield? Why not? W. P. Kinsella imagined that very *Dream Field* and made it a centerpiece in his short story *Shoeless Joe Comes to Iowa*. With the support of editor Larry Kessenich, Kinsella's follow-on novel *Shoeless Joe* became an award-winning best-seller. Screenwriter-director Phil Robinson envisioned that novel becoming a feature film and brought it to the silver screen as *Field of Dreams* – both a critical and popular success. Lead actor Kevin Costner sensed something special about the movie set and suggested to farm owner Don Lansing that he not destroy the field. The *Field of Dreams Movie Site* still stands in that Dyersville cornfield with tens of thousands flocking there every year to celebrate the film, the game, and the bonds of family. Finally, thanks to the vision of baseball entrepreneurs Denise Stillman and Tom Mietzel, Major League Baseball is scheduled to play a regular season game in a second cornfield ballpark constructed adjacent to the original film field.

I daresay that nothing beyond *Shoeless Joe Comes to Iowa* could have been foreseen by Bill Kinsella. Some might fantasize a cosmic plan behind all of these convoluted goings on. But, what we have here is life proceeding as it always does – with context, consequence, and coincidence plying their trade – and historians doing their level best to make sense of it all.

If we accept that Kinsella was utterly blind to what the future held for his original short story, then we can ponder one more vision of what might yet become of Kinsella's *Dream Field*. With the novel, the film, and the active movie site in his rearview mirror, Kinsella teases us with a titillating prophesy, "*Field of Dreams* the musical is out there in the cosmos, ethereal as *Brigadoon*, lurking, waiting patiently, being groomed for the stages of the world" (2015). Could Kinsella's stunning prediction come to pass? Never say never!

To conclude this vignette, Fred Mason's 2018 essay on the dreamscape of *Shoeless Joe* offers a special window into Kinsella's own storytelling and on those who would mythologize the author himself, "Ray's little cornfield ballpark in Iowa has become mythic itself. This is appropriate, since magic, dreaming, and myth are at the heart of W. P. Kinsella's *Shoeless Joe*. In allowing all of the characters in the novel to have their dreams fulfilled, Kinsella offers baseball (and Iowa) as a social world for dreams and dreamers" (22).

It is therefore altogether fitting that in the *Field of Dreams* film, when Shoeless Joe Jackson materializes after Ray Kinsella builds the cornfield baseball diamond, Jackson asks, "Hey, is this heaven?" Kinsella replies, "No, it's Iowa."

Fun fact: Kevin Costner's film *Revenge* made less than $16 million at the box office, whereas *Field of Dreams* made more than $84 million!

REFERENCES

Costner, K. and Costas, B. (2014, June 13–15). *Field of Dreams* 25 Years Later. www.youtube.com/watch?v=UONw_p89rBM

Easton, N. J. (1989, April 21). Diamonds Are Forever: Director Fields the Lost Hopes of Adolescence. *Los Angeles Times*. www.latimes.com/archives/la-xpm-1989-04-21-ca-2279-story.html

Kinsella, W. P. (1980). *Shoeless Joe Jackson Comes to Iowa*. Ottawa: Oberon.

Kinsella, W. P. (1982). *Shoeless Joe*. Westminster, MD: Ballantine Books.

Kinsella, W. P. (2004, March 1). State of Dreams: The Author's Novel Spawned a Landmark Film about Magic amid the Cornfields. *Sports Illustrated*.

Kinsella, W. P. (2015). *The Essential W. P. Kinsella*. San Francisco, CA: Tachyon Publications.

Mason, F. (2018). W.P. Kinsella's *Shoeless Joe*: The Fairy Tale, the Hero's Quest, and the Magic Realism of Baseball. In A. Abdou and J. Dopp (eds.), Athabasca,

Alberta: University of Athabasca Press. *Writing the Body in Motion: A Critical Anthology on Canadian Sport Literature*, 1–14. https://doi.org/10.15215/aupress/9781771992282.01

Okrent, D. (1982, July 25). Imaginary Baseball. *The New York Times.* www.nytimes.com/1982/07/25/books/imaginary-baseball.html

O'Leary, J. (2020, June). If You Write It: The University of Iowa Author Who Inspired the *Field of Dreams. Iowa Magazine.*

Reichard, K. (2020, August 10). With Game Postponed, *Field of Dreams* Lies Fallow. *Ballpark Digest.* https://ballparkdigest.com/2020/08/10/with-game-postponed-field-of-dreams-lies-fallow/

Steele, W. (2019). *Going the Distance: The Life and Works of W. P. Kinsella.* Madeira Park, British Columbia: Douglas & McIntyre.

SECTION 3 **Putting It Together**

26 Context, Consequence, and Coincidence

The preceding collection of vignettes contains a rich assortment of individual cases, each exemplifying important behavioral innovations. With so much variation, it would seem to be an impossible task to identify any underlying commonalities among them. Nevertheless, in a good faith effort to do just that, I earlier asked readers to consider the roles played by the *Three Cs*: context, consequence, and coincidence. These factors seem to be prominent in most, if not all, of the various vignettes. So, as a test case, let's step through the Fosbury Flop and see how well that general approach fares.

Context. Dick Fosbury was a lanky high schooler with so-so high jumping skills. He was extremely uncomfortable adjusting to the "belly roll" technique that his coach strongly recommended; despite endless efforts, he just never managed to get the hang of it. Largely out of desperation, Fosbury reverted to the less fashionable, but at least for him, more comfortable "scissors" style.

Consequence. Extemporaneous trial-and-error experimentation quite inadvertently spawned Fosbury's highly unorthodox and original technique of sailing backward over the high bar – with Fosbury now surpassing his own best performance. Countless failures had finally yielded surprising success. Fosbury stuck with his freshly minted "flopping" technique all the way to the 1968 Olympic Games in Mexico City and to a Gold Medal.

Coincidence. In his junior year, Fosbury's high school installed foam rubber in its high jump landing pit; it was the first high school in the state of Oregon to do so. Because the Flop finishes with the jumper precariously falling on the back of the neck and shoulders, this compressible material could safely cushion Fosbury's fall, which would otherwise have been far more dangerous if only sand or sawdust had greeted him.

I believe that this overall analysis can be fittingly applied to those vignettes for which substantial background information is available. I invite readers to judge for themselves. Nonetheless, critics might contend that my analysis is too simplistic. Isn't there much more to say about behavioral invention than deconstructing what is a complex process into the *Three Cs*? I'm entirely open to that possibility. The Fosbury Flop provides a good starting point for just such an expanded and nuanced discussion. So, let's keep at it.

FURTHER DEVELOPING THE THREE CS: CONTEXT

When I began seriously contemplating the nature of behavioral innovation, I immediately recalled the Fosbury Flop. His famous Olympic jump had happened during the summer following my graduation from UCLA. As were so many others, I was flabbergasted by Fosbury's novel and idiosyncratic technique. I wondered how he could ever have invented this wildly eccentric maneuver. I concluded that the Fosbury Flop would never have arisen had he alone not stumbled upon it.

I remained firm in my conviction until many years later when I began more thoroughly researching the matter. I now concede that I had been flat wrong! The backward leap was no "one-off." Not one, but three young athletes are now recognized as having developed the backward leap without knowing of anyone who had done so before.

Let's proceed chronologically. Bruce Quande was born in 1945 and lived in Kalispell, Montana. Dick Fosbury was born in 1947 and lived in Medford, Oregon. And, Debby Brill was born in 1953 and lived in Aldergrove, British Columbia. Eight years separated their births and between 550 and 800 miles separated their hometowns. There is no evidence that any of them knew of the others' flopping technique while they were experimenting with their own individual versions. Independent invention thus seems assured.

I've previously discussed in detail Fosbury's fascinating story of behavioral invention and noted how little we know about Quande's story. But, it's the richness of Brill's story that has recently forced me to abandon my original belief that only Fosbury could have invented the Flop.

Fosbury began high jumping at ten or eleven years of age. Debbie Brill also began high jumping in elementary school. As did Fosbury, the gangly preteen girl also found the "belly roll" unnatural and began experimenting with the "scissors" technique. Her experimentation was initially hindered by the inhospitable sand and sawdust landing pits at her school. However, Brill's strongly supportive father accelerated her progress by building a landing pit on the family farm and outfitting it with a cushion made of foam rubber pieces encased in a fishing net (more about the Brill Bend can be found in Hutchins, 2014; Russell, 2017; Verschoth, 1971).

Brill developed the reverse jumping style without any direction from her father or her coaches. In Brill's own words from a revealing interview recently reported by Russell (2017, [all emphasis added]), "Back then, sport was not as mainstream as it is today, and *we had to make it up as we went along*. I was left to my own devices, and part of my talent was that I was able to be completely *in the moment* of the jump." As to the form of the jump, Brill explained, "It was just a *natural* extension of what my body was telling me to do. It was *physical intuition*. It wasn't anything taught." Of course, it didn't arise full blown. "Originally, it was sort of a sideways thing; it later developed into the full backward jump." In this way, the Brill Bend was born.

Brill's detailed and engaging story is strikingly similar to Fosbury's! Each tale testifies to the entirely inadvertent discovery of what Hoffer (2009) had colorfully called an outrageous perversion in form and technique of all acceptable methods of jumping over obstacles – a style unto itself. How then are we to understand not only the similar *product* of Fosbury's and Brill's pioneering efforts but also the similar *process* of its origin and evolution?

This question takes us to the next step in the progression of my own thinking; it came in the course of writing a review

(Wasserman, 2012) of a captivating book by Jonnie Hughes (2011), *On the Origin of Tepees*. Hughes's main thesis was that if evolution by natural selection explains the origin of the human species, then does the parallel process of selection by consequences also explain the origin of what we humans do and make? His answer was "yes." The Fosbury Flop and the Brill Bend fit perfectly within this evolutionary framework – the Law of Effect rather than the Law of Natural Selection now playing the prime functional part (also see Blumberg and Wasserman, 1995; Skinner, 1966).

Consider a second and far more famous example: Charles Darwin's own revolutionary theory of evolution through natural selection. Hughes observed that Darwin's theory did not abruptly or miraculously flash into his mind. Rather, it had a lengthy and tortuous gestation period, unfolding throughout most of Darwin's adult life. Moreover, Darwin's theory was not hatched fully fledged; it was extensively revised in six editions of his *On the Origin of Species*, from 1859 to 1872.

Hughes believed that this case of Darwin and his theory of evolution by natural selection qualified as an illustration of "human genius." To Hughes, "genius" represented the rare situation in which the evolution of an idea takes place in just one mind rather than in many.

Surprisingly, Hughes did not discuss another parallel evolutionary theory authored by fellow British scientist Alfred Russell Wallace (1823–1913). Darwin and Wallace developed their theories at much the same time, each having been engaged as naturalists and collectors of thousands of plants and animals in their youthful ocean voyages.

But, Darwin and Wallace did not devise simply similar theories. According to Gross (2010), "in the most important development in the history of biology and one of the most important in the history of ideas, they came to an essentially identical view of the origin of species by natural selection of inherited variations" (505). In fact, at a meeting of the Linnaean Society in London on July 1, 1858, Darwin and Wallace presented back-to-back papers describing their independently devised theories of natural selection. One might thus say that both Darwin and Wallace were "geniuses" who happened to have the same brilliant idea.

Here, then, are two well-documented cases of independent innovation: Darwin and Wallace formulating the Law of Natural Selection, and Fosbury and Brill originating the Flop or Bend (respecting proper alliteration, of course). In the words of the inimitable Yogi Berra, it's déjà vu all over again!

Renowned American sociologist of science Robert Merton (1968) reflected on the significance of such multiple independent discoveries, calling particular attention to the reputational inequality that often results. The terms "Fosbury Flop" and "Darwinian evolution" prove Merton's point – with Brill and Wallace receiving little or no credit.

Cep (2019) has recently revisited Merton's "multiples," stressing the critical role played by what I've called the prevailing context in the simultaneity of these discoveries, "The problems of the age attract the problem solvers of the age, all of whom work more or less within the same constraints and avail themselves of the same existing theories and technologies."

The implications of Cep's proposal are indeed weighty. They take us back to Hughes's 2011 book and to his most thought-provoking proposal. Regarding the Law of Natural Selection, Hughes claimed that, "Darwin didn't have a brilliant Idea; the Idea had a brilliant Darwin" (140). If that puzzling proposal is to be accepted, then so too must this one, "Wallace didn't have a brilliant Idea; the Idea had a brilliant Wallace." And, more comprehensively, the Idea of natural selection of inherited variations had *both* a brilliant Darwin and a brilliant Wallace.

Proceeding with the same logic, let's also ask: What about the backward high jump? Here, too, we're invited to believe that this singularly rebellious Idea somehow existed outside of its two most accomplished and recognized originators: a brilliant Fosbury and a brilliant Brill.

What an outlandish proposal! Innovative ideas and behaviors *having* brilliant people? And, yet, consider Fosbury's own assertion, "I believe that the flop was a *natural* style and I was just the first to find it. I can say that because the Canadian jumper, Debbie Brill was a few years younger than I was and also developed the same technique, only a few years after me and without ever having seen me." Brill too stressed that this utterly uncommon style was completely "natural" and based on "physical intuition," which just happened "in the moment."

Fine, you say. Maybe there's simply something special about the invention of abstract scientific theories or human athletic actions that encourages this strange kind of speculation.

Let's now examine the views of one more innovator – indeed, the most famous inventor of *things* – Thomas Alva Edison (1847–1931). Consider Edison's astounding assertion, what Edmund Morris (2019) deemed to be the frankest of all Edison's many self-appraisals,

> I never had an idea in my life. I've got no imagination. I never dream. My so-called inventions already existed in the environment – I took them out. I've created nothing. Nobody does. There's no such thing as an idea being brain-born; everything comes from the outside. The industrious one coaxes it from the environment.

Unpacking this jam-packed claim is a formidable challenge. Thompson (2019) offers us some assistance: "This [claim] can be read in several ways – as provocative overstatement, as an honest description of creativity's mechanics, or as a paean to the inventor's workaholism." Provocative? Undeniably. A song of praise to effort and perseverance? Absolutely. After all, Edison famously insisted that "Genius is one per cent inspiration, ninety-nine per cent perspiration." A proper characterization of the processes of creativity? I'm not yet convinced that ideas reside outside the body. Edison's "externalist" proposal must be the most radical I've ever encountered and one that remains open for debate.

FURTHER DEVELOPING THE THREE CS: COINCIDENCE

We previously noted that the cushioned landing pit may have provided an especially supportive context for development of the backward or reverse high jump. Perhaps by chance, both Dick Fosbury and Debbie Brill had early access to the novel landing cushion. Of course, we'll never know just how important this factor was for the high jumping innovation that Fosbury and Brill pioneered. Nor will we know whether or not other high jumpers would have devised the technique if neither Fosbury nor Brill had done so. Greater clarity on the role of coincidence in behavioral creativity comes from cases of individual happenstance.

The most unequivocal examples of coincidence are those that are unique to an individual innovator; as such, they are not subject to being subsumed by the broad context in which innovations arise. Many vignettes in this book contain such instances. Here's a sampling,

> While out riding, Tod Sloan stopped his mount's sudden bolting by rising from the saddle and climbing forward onto his horse's neck. Sloan noticed that doing so freed the steed's stride, prompting him to foster the "monkey crouch" style of race riding.
>
> During one of his many maladies, Ansel Adams was gifted a book from his Aunt Mary, *In the Heart of the Sierras*, written by James Mason Hutchings and containing photographs of Yosemite by George Fiske. Ansel became enchanted by Yosemite and photography, later making these the twin passions of his adult life.
>
> Florence Nightingale happened to meet Sidney Herbert in Rome through mutual friends: Charles and Selina Bracebridge. This was later to bring together the two heroes responsible for reforming British military care following the Crimean War.
>
> Basil Twist often foraged for interesting *objets d'art* that were deposited in recycling piles along the streets of New York. One day, he chanced to spot a cracked fish tank, which he proceeded to repair. Later, Twist happened to eye a recording of Berlioz's *Symphonie Fantastique*, which he recalled having been in his parents' collection. These two accidental discoveries paved the way to one McArthur "Genius Grant."

While working in Mexico after fleeing war-torn Spain, Ignacio Ponseti sought the counsel of a physician, Juan Faril, who happened to have clubfoot disorder. Dr. Faril highly recommended Ponseti to Arthur Steindler who supported Ponseti's appointment to The University of Iowa medical school. Steindler's first project for young Ponseti: analyze data on clubfoot surgery.

Coincidences such as these are said to be fortuitous or serendipitous (Merton and Barber, 2004). Their pursuit can lead to important innovations. But, so too can misfortunes. Consider these,

Jockey Jack Westrope suffered an injury to his left leg in a starting gate accident that prevented him from fully flexing his left knee. This meant that, when racing, Westrope's left leg had to be more extended than his right, thereby requiring him to place the stirrups at notably different heights. Because races in the United States are run in a counterclockwise manner, this particular uneven positioning on the thoroughbred actually permitted Westrope to lean more effectively into left turns. The fact that other jockeys eagerly copied Westrope's acey-deucy riding style was most assuredly no accident.

Jakob Kolletschka's untimely death stunned fellow physician Ignaz Semmelweis. But, the cause of death – puerperal infection caused by an errant scalpel pricking Kolletschka's finger – provided Semmelweis with the critical clue for solving the mystery surrounding the death of so many women from childbed fever in the First Maternity Division of the Vienna Maternity Hospital.

Despite his fervent efforts, Francis Brackin – the Cambridge lawyer who argued the 1612 lawsuit on behalf of Mayor Thomas Smart against the University – lost the case. Yet, that indignity did not satisfy University Teaching Fellow George Ruggle's quest for revenge. Exacting further retribution, Ruggle subsequently authored the satirical play *Ignoramus,* basing its key character, Ambidexter Ignoramus, on the insolent and arrogant Brackin. So, for the last 400 years, the word "ignoramus" has denoted foolish and ignorant people like Francis Brackin.

A Sunday afternoon meal suddenly went horribly wrong. Irene Bogachus slumped forward at her dinner table and began to turn blue, her life now in jeopardy. Irene's husband, Edward, raced to a neighbor's home and returned with Isaac Piha. Having just the night before read a newspaper story describing a new method of expelling food from a choking victim, Piha successfully used it to eject a large chunk of chicken from Irene Bogachus' throat. Never before had Henry Heimlich's Maneuver been used to save a person from choking.

FURTHER DEVELOPING THE THREE CS: CONSEQUENCE

Having further discussed the roles that context and coincidence play in behavioral innovation, it's now time for us to consider in greater depth the consequences of our actions. Specifically, we must give greater thought to the Law of Effect and its relation to the Law of Natural Selection. These two selectionist laws are critical to how humans and other animals adapt to their worlds (Rosenbaum, 2014; Stahlman and Leising, 2018).

The Law of Natural Selection and the Law of Effect

In his autobiography, Charles Darwin forcefully denied that an intelligent designer was responsible for the origin and diversity of life on earth. Instead, he proposed that a mindless, mechanical law of nature – the Law of Natural Selection – underlies organic evolution. Darwin (1887/1958) wrote, "There seems to be no more design in the variability of organic beings and in the action of natural selection, than in the course which the wind blows. Everything in nature is the result of fixed laws" (87). The exquisitely adaptive construction of humans and other life forms thus represents apparent design without a designer.

Paradoxically, in this same work, Darwin quite uncritically accepted intelligent design in the construction of exquisitely contrived devices by humans. "The old argument of design in nature ... which formerly seemed to me so conclusive, fails, now that the law of natural selection has been discovered. We can no longer argue that, for instance, the beautiful hinge of a bivalve shell must have been made by an intelligent being, like the hinge of a door by man" (87). Darwin never expanded on this claim, so we don't know precisely how he'd have believed humans came to make a door hinge or countless other contrivances. But, his connecting the construction of a door hinge with human intelligence underscores a profound difference *in kind* between nature's hinge and humans' hinge.

Still, what do we actually know about the invention of hinges and – given the earlier presented vignette – musical instruments like violins? Must we believe that people fashioned these devices with full foresight of their later shape and operation as implied by the notion of intelligent design? Or might there be another less mysterious way to explain their development? Here are some tentative answers to these challenging questions.

First, we don't really know how humans came to make either hinges or violins; the historical record is woefully incomplete and inconclusive, although the studies of Chitwood (2014) and Nia et al. (2015) have provided new data and offered fresh insights into our understanding of violin evolution. What we do know about the origin of human contrivances and behaviors – from the origin of tepees, light bulbs, and telescopes to the dance moves of entertainers, the high jumping styles of Olympic athletes, and the riding stances of thoroughbred jockeys – seriously questions the necessity of intelligent, premeditated design in the process (Hughes, 2011; Petroski, 2012; Ridley, 2020; Simonton, 2012, 2015; Wasserman, 2012; Wasserman and Blumberg, 2010). Trial-and-error learning and sheer chance often turn out to be the particularly important contributors to developing the things we make and do.

Second, the notion of premeditated, intelligent design may be little more than a romantic fiction that will turn out to be as useless in explaining the origin of novel inventions and behaviors as it has proven to be in illuminating the origin of species. In the face of sound scientific evidence, shouldn't we be just as willing to jettison intelligent design as an explanation for the origin of human inventions as Darwin was to jettison intelligent design as an explanation for the origin of human beings?

Third, we already have a good and ever-growing understanding of how most human behaviors and inventions are created. They develop via another fixed law of nature, which generally parallels Darwinian selection – the Law of Effect (Dennett, 1975; Skinner, 1966, 1969, 1974). According to the Law of Effect, out of a diverse repertoire of behaviors, those that are followed by reinforcers tend to be repeated and to predominate, thereby promoting increasingly adaptive action. This mechanical, trial-and-error process is capable of supporting novel behaviors and inventions akin to the way the Law of Natural Selection produces novel organisms. Skinner (1966) called this process *selection by consequences* to stress the analogy between behavioral selection (operating within the lifetime of an individual organism) and natural selection (operating across the lifetimes of many organisms).

The Law of Effect: Asserting Its Role in the Evolution of Behavior

How do species originate? Darwin daringly departed from a biblical account and made a bold substitution: He replaced an all-intelligent *deity* with a natural selective process capable of producing organic change. Skinner (1974) offered a different but equally bold interpretive substitution: He replaced an intelligent human *mind* with a second selective process – the shaping of adaptive behavior by contingencies of reinforcement.

Skinner's (1974) perspective held that the contingencies of survival and the contingencies of reinforcement can each produce novel and adaptive outcomes, "Natural selection explained the origination of millions of different species ... without appealing to a creative mind" (224). In the case of behavior, "contingencies of reinforcement may explain a work of art or a solution to a problem in mathematics ... without appealing to a different kind of creative mind or to a trait of creativity" (224).

Summing up, Skinner (1974) proposed this parallel, "As accidental *traits*, arising from mutations, are selected by their contribution to survival, so accidental variations in *behavior* are selected by their reinforcing consequences" (114 [all emphasis added]). This parallel is profound and it underscores how selectionist principles can explain both the origin of species and the origin of acquired adaptive behaviors; nonetheless, quite different biological mechanisms may lie at the root of natural selection (genetics) and selection by consequences (neurobiology).

Richard Dawkins (1986) has contended that people's continuing reluctance to accept Darwinian evolution results from our species' remarkable success in creating innumerable useful things; other than intelligent design, how else can we explain the invention of the automobile, the computer, and the air conditioner? Given the amazing utility of such biological creations as the hand, the eye, and the brain, "it took a very large leap of the imagination for Darwin ... to see ... a far more plausible way [than a divine intelligence] for complex 'design' to arise out of primeval simplicity. A leap of imagination so large that, to this day, many people seem still unwilling to make it" (xii).

Are we not long overdue for taking another very large leap of imagination? The creative and seemingly well-designed things we do may also arise out of primeval simplicity: namely, the shaping of behavior by the mechanical process of trial and error. Perhaps the Law of Effect should finally be placed directly alongside the Law of Natural Selection. Together, these two basic selectionist principles can produce organisms that are exquisitely adapted to their surroundings, and do so according to laws that are entirely natural, mechanical, and operate without premeditation. This framework would at long last represent the "tidy unification of the sciences of life and behavior" (Stahlman and Leising, 2018, 926).

SOME POSSIBLE CONCERNS

I'm sure that many readers will have concerns about my point of view. So, let me round out this section with responses to some of those anticipated concerns.

Overt or Covert Processes in Behavioral Evolution?

One concern with my approach is that a purely behavioral account based on the Law of Effect misses the possibly pivotal role played by covert cognitive processes in the creation of novel behaviors or contrivances (Rosenbaum, 2014, makes the clever counterclaim that cognitive processes themselves comport with the mechanics of the Law of Effect). Considering the substantial time, effort, and expense involved in violin construction, for example, is it not likely that master luthiers made every effort to behave rationally and economically, generating novel models and selectively retaining only those that could be foreseen to be acoustically superior – and doing so entirely in the theater of their consciousness?

The possible adaptive significance of such premeditated or intelligent designing was proposed by Karl Popper, one of the twentieth century's most influential philosophers of science. Popper (1978) maintained that our *ideas* are more expendable than *ourselves*, "Let our conjectures, our theories, die in our stead" (354). Dennett (1996) deemed Popper's proposal to be so compelling that he dubbed those organisms that were capable of such *virtual* trial-and-error *Popperian* creatures; in contrast, he dubbed those organisms whose behaviors were modifiable solely by *experienced* contingencies of reinforcement *Skinnerian* creatures.

There are at least two ways to respond to this possible limitation of a behavioral account. First, at the *empirical* level, given the detailed observations of violin construction made by Nia et al. (2015), it appears that particularly small sound-hole variations actually distinguished violins made over many decades of fabrication; indeed, across some 200 years (1560–1750), sound-hole length increased by only 15 mm. Such small, intergenerational variations are not at all what would have been expected if innovative models envisioned in the "mind's eye" of master luthiers had inspired longer sound-holes; considerably greater and more abrupt changes in f-hole length would have been far faster and less costly to produce. Those slight sound-hole changes are also likely to have occurred in the midst of many other changes in violin fabrication, any of which might also have promoted greater acoustic power. Those other changes ought to have made the cognitive chore of pinpointing the effective physical ingredient of enhanced violin performance exceptionally difficult to discern – even with the keenest vision of one's "mind's eye."

Second, at the *theoretical* level, Dennett's distinction between Skinnerian and Popperian creatures may be more apparent than real. Skinner (1969) himself proposed that, "An adequate science of behavior must consider events taking place within the skin of the organism, not as physiological mediators of behavior, but as part of behavior itself. It can deal with these events without assuming that they have any special nature or must be known in any special way. *The skin is not that important as a boundary* [emphasis added]. Private and public events have the same kinds of physical dimensions" (228).

The claim that the skin does not represent a critical boundary to behavioral analysis is, most assuredly, highly controversial (Baum, 2011; Catania, 2011). What cannot be disputed, however, is that – whether inside or outside the skin – the basic elements of behavior analysis (stimulus, response, reinforcer) can control the behavior of organisms without the participation of foresight. Skinner's dismissal of the skin as a boundary – a purely theoretical claim for which decisive empirical evidence is lacking – need not represent an invitation to premeditation in the creation of novel behaviors or devices.

What Exactly Is Meant by Design?

The word *design* has acquired many meanings over many centuries. Remarkably, by most standard definitions, we are forced to conclude that the highly original and adaptive behaviors detailed in the preceding vignettes were *not* designed. Even the one physical object that we considered – the violin – cannot properly be said to have been designed. Of course, generations of luthiers crafted violins over hundreds of years. But, their doing so did not accord with today's hallmarks of design.

Consider these two definitions of the verb *design*. (1) To create, fashion, or construct something according to a sketch, outline, pattern, or plan. But, the world's most treasured violins of the seventeenth and eighteenth centuries (or their less revered forerunners) were not constructed according to detailed blueprints or computer renderings, as might now be the case. Furthermore, the evolution of the violin would have required the planning of *future* variants that differed from the *prevailing* versions; how could those future designs have been foreseen or passed on to unborn luthiers without mechanical drawings? (2) To form or conceive of something in the mind. We certainly have no convincing scientific evidence to suggest that such mental processes participated in the evolution of the violin.

People do undoubtedly build many things: violins, hinges, bridges, computers, and airplanes. How we as a species have come to do so is both a matter of process and history. Dawkins (1986) has noted that we have become quite familiar with a world that is dominated by amazing feats of engineering. But, that very familiarity leads to a lamentable brand of intellectual complacency when we earnestly endeavor to understand just how those feats came to pass.

Consider the case of people today building an airplane. Here's how Dawkins describes the process,

> However incompletely we understand how an airliner works, we all understand by what general process it came into existence. It was designed by humans on drawing boards. Then other humans made the bits from the drawings, then lots more humans (with the aid of other machines designed by humans) screwed, rivetted, welded or glued the bits together, each in its right place. The process by which an airliner came into existence is not fundamentally mysterious to us because humans built it. The systematic putting together of parts to a purposeful design is something we know and understand. (3)

Anyone who is even remotely familiar with the history of heavier-than-air flight will appreciate how useless this account is if we are at all interested in properly explaining the *evolution* of the airplane; Dawkins' account curiously begins with how airplanes are *now* constructed rather than with their *origin*. Nor will this kind of jejune, ahistorical description be in any way useful if we wish to explain the origin and evolution of the violin, or any other device for that matter.

The true story behind the development of violins and airplanes requires an understanding of the *behavior* of luthiers and aeronautical engineers. That is precisely why the evolution of even things like violins and airplanes is a matter that demands the expertise of *behavioral scientists*; they alone are in the position of weaving the kinds of physical evidence that have been so meticulously collected and physically analyzed by Chitwood (2014) and Nia et al. (2015) into a coherent theoretical narrative that can properly explain the evolution of the many remarkable things that we humans make and do.

How Compelling Is My Case?

At the outset, I advised readers that my case against insight and foresight in the origin of behavioral innovations was not going to concentrate on dull and detailed laboratory findings; instead, I was going to focus on real life examples in the hope of telling a convincing story. Through this approach, I hoped to advance progress in the scientific understanding of creativity – not simply by telling *another* story, but a *better* story. I fully understand that this narrative tactic has its downsides. Let's consider them now.

Merton (1945) expressed serious misgivings about explanations that are offered after the evidence is already in hand, as was necessarily the case with the twenty-five vignettes that I presented. Merton called these *post factum* (after the fact) explanations. One complaint that can't be leveled against such explanations is that they conflict with the evidence; that's because they have to accord with the reported observations in order to be taken at all seriously. The real complaint is that other rival interpretations may just as effectively explain the facts at issue. So, a *post factum* explanation may merely be *plausible* rather than *compelling*.

This is an entirely valid concern. In my defense, I would suggest that the sheer variety of different innovations I have considered – from medical discoveries and therapies to innovations in athletic competitions to colorful cultural celebrations to arcane political practices – should enhance the plausibility of my behavioral science account over rival interpretations.

Then, there's the matter of my having "cherry picked" these particular vignettes. Perhaps the stories I recounted were selected solely to prove my point. If I had chosen other vignettes, then they might have laid waste to my pet theory.

This is another valid concern. In my defense, I would note that, for the past fifteen years, I've been assiduously searching for cases of behavioral innovation that provide sufficient historical information to assess the plausibility of my natural science account. I've reported twenty-five such cases in this book. I have many more in my files that I hope to share in the future. I welcome hearing from readers about instances where insight and foresight clearly undermine my current interpretation. I don't claim to be all knowing and I'm fully prepared to concede mistakes.

In that spirit, I'd like to examine one final story that I'm aware has famously been claimed to convincingly confirm the role of sudden insight in scientific discovery. It's the case of Kekulé's discovery of benzene's chemical structure. Benzene is the most commonly encountered aromatic organic compound. The structural representation for benzene involves a six-carbon ring – represented by a hexagon. Friedrich August Kekulé (1829–1896) was a German organic chemist who is duly credited with creating that representation.

Kekulé's 1865 report of that breakthrough was so significant that twenty-five years later, in 1890, the German Chemical Society held a special celebration in his honor. Kekulé took that occasion to describe a key episode in the process of his innovation. "Kekulé's account of circumstances leading to this discovery has been the single most cited personal report in psychological writings on creativity" (Rothenberg, 1995, 419). Here is the text of his remarks (in its full translation by Leslie Willson from the paper by Rothenberg, 1995, 424–425) which had been preceded by his telling of an earlier "visionary" episode,

> During my sojourn in Ghent in Belgium I occupied an elegant bachelor apartment on the main street. My study, however, was located in a narrow side lane and during the day had no light. For a chemist, who spends daylight hours in the laboratory, this was no disadvantage. There I sat, writing on my textbook; but it wasn't going right; my mind was on other things. I turned the chair to face the fireplace and slipped into a languorous state. Again atoms fluttered before my eyes. Smaller groups stayed mostly in the background this time. My mind's eye, sharpened by repeated visions of this sort, now distinguished larger figures in manifold shapes. Long rows, frequently linked more densely; everything in motion, winding and turning like snakes. And lo, what was that? One of the snakes grabbed its own tail and the image whirled mockingly before my eyes. I came to my senses as though struck by lightning; ... I spent the rest of the night working out the results of my hypothesis.

What are we to make of Kekulé's vibrant story? Despite its popularity across many generations of scholars, some authors, particularly Rudofsky and Wotiz (1988), doubt that the "snake-dream" was truly the insightful inspiration behind revealing the structure of benzene. Indeed, they contend that little would be lost by deeming this story simply to be an engaging anecdote that Kekulé told to entertain his admiring audience at the 1890 Benzol Fest.

Rocke (1985) is more inclined to accept Kekulé's story at face value. However, Rocke's extensive archival research puts the famed benzene story into a strikingly different light – one that views Kekulé's pioneering work as continuous and evolutionary rather than discontinuous and revolutionary.

Rocke's (1985) analysis starts with the conventional story line, "Kekulé's theory has often been depicted as one of those anomalous cases where a sudden inspiration, even perhaps as the result of a dream or hallucination, dramatically transformed a scientific field" (377). Rocke then proceeds methodically and meticulously to correct the record, "Contrary to most accounts, and to the implication of the dream anecdote told out of context, it is now clear that the Benzene theory did not fall into Kekulé's half-awake mind fully formed – or even partially formed. It was at most the ring *concept* that arrived by this semiconscious or unconscious process, a concept which ... was not without precedent" (377).

After detailing all of the many precedents that paved the way for Kekulé's discovery as well as the critical revisions and elaborations Kekulé made after the dream incident had taken place, Rocke offers this final verdict, "The theory itself was developed only slowly, one might even say painfully, over the course of several years, before its first codification" (377) – the dream episode clearly representing a single step in a long and winding road of hypotheses and experiments.

Some twenty-five years after writing this critical analysis of Kekulé's discovery, Rocke (2010) offered an even broader perspective on the role of visualization in science – and especially in chemistry. Rocke's sense is that "human minds work far more visually, and less purely linguistically, than we realize" (xii). Rather than being an outlier, Kekulé might be highly representative of scientists whose job involves "pondering the invisible microworld" (xiii).

To support his claim, Rocke (2010) called on Kekulé's additional recollections about the discovery of benzene's structure (see 65 for the following quotations). These recollections nicely elucidate the nature of the process, stressing the prevailing theoretical *zeitgeist* as well as Kekulé's own educational history – both contributing to what I have called the *context* of innovation. At that 1890 celebration, Kekulé remarked that "[my eclectic education] and the direction that my early architectural education provided to my mind, gave me an irresistible need for visualizability; these are apparently the reasons that twenty-five years ago it was in my mind ... where those chemical seeds of thought that were floating in the air found appropriate nourishment." Two years later, Kekulé said of those "floating" thoughts that "sooner or later they would have been expressed ... perhaps in a different way than I did. [But] it would have become merely a 'paper chemistry,' only the architect was able to provide the arrangements of atoms a living, spatial conception." Kekulé was obviously the architect who beat others to the proverbial punch.

So, there you have it. The single most salient and cited story about the role of insight in scientific discovery collapses when subjected to thorough historical analysis.

Please appreciate that I specifically chose to end this section of my book with an account of Kekulé's famous "snake" story because I fully expected it to serve as a *counterpoint* to the position I was advocating. So, I was quite surprised to learn that Rocke's assiduous analysis so closely accorded with my own view. I'm thus reassured by this reassessment of Kekulé's "insightful" reverie that my perspective does have merit.

REFERENCES

Baum, W. M. (2011). Behaviorism, Private Events, and the Molar View of Behavior. *The Behavior Analyst, 34,* 185–200.

Blumberg, M. S. and Wasserman, E. A. (1995). Animal Mind and the Argument from Design. *American Psychologist, 50,* 133–144.

Catania, A. C. (2011). On Baum's Public Claim That He Has No Significant Private Events. *The Behavior Analyst, 34,* 227–236.

Cep, C. (2019, October 21). The Real Nature of Thomas Edison's Genius. *The New Yorker.*

Chitwood, D. H. (2014). Imitation, Genetic Lineages, and Time Influenced the Morphological Evolution of the Violin. *PLOS ONE, 9,* e109229.

Dawkins, R. (1986). *The Blind Watchmaker.* New York: Norton.

Darwin, C. R. (1958). *The Autobiography of Charles Darwin 1809–1882.* London: Collins. (Original work published 1887)

Dennett, D. (1975). Why the Law of Effect Will Not Go Away. *Journal for the Theory of Social Behaviour, 5,* 169–187.

Dennett, D. (1996). *Kinds of Minds.* New York: Basic Books.

Gross, C. (2010). Alfred Russell Wallace and the Evolution of the Human Mind. *The Neuroscientist, 16,* 496–507.

Hofer, R. (2009, September 14). The Revolutionary. *Sports Illustrated.*

Hughes, J. (2011). *On the Origin of Tepees.* New York: Free Press.

Hutchins, A. (2014, June 28). 'When I First Started, I Was All Gangly and Awkward' – Debbie Brill Figured out How to Jump Higher. *Maclean's.* www.macleans.ca/news/canada/record-setting-high-jumper-who-invented-technique/

Merton, R. K. (1968). The Matthew Effect in Science. *Science, 159,* 56–63.

Merton, R. K. (1945). Sociological Theory. *American Journal of Sociology, 50,* 462–473.

Merton, R. K. and Barber, E. (2004). *The Travels and Adventures of Serendipity.* Princeton, NJ: Princeton University Press.

Morris, E. (2019). *Edison.* New York: Random House.

Nia, H. T., Jain, A. D., Liu, Y., Alam, M-R, Barnas, R., and Makris, N. C. (2015). The Evolution of Air Resonance Power Efficiency in the Violin and Its Ancestors. *Proceedings of the Royal Society A, 471,* 20140905.

Petroski, H. (2012). *To Forgive Design.* Harvard, MA: Belknap Press.

Popper, K. (1978). Natural Selection and the Emergence of Mind. *Dialectica, 32,* 339–355.

Ridley, M. (2020). *How Innovation Works.* New York: Harper.

Rocke, A. J. (1985). Hypothesis and Experiment in the Early Development of Kekulé's Benzene Theory. *Annals of Science, 42,* 355–381.

Rocke, A. J. (2010). *Image and Reality: Kekulé, Kopp, and the Scientific Imagination.* Chicago, IL: University of Chicago Press.

Rosenbaum, D. A. (2014). *It's a Jungle in There.* New York: Oxford University Press.

Rothenberg, A. (1995). Creative Cognitive Processes in Kekulé's Discovery of the Structure of the Benzene Molecule. *American Journal of Psychology, 108,* 419–438.

Rudofsky, S. F. and Wotiz, J. H. (1988). Psychologists and the Dream Accounts of August Kekulé. *Ambix, 35,* 31–38.

Russell, S. (2017, July 7). Debbie Brill Raised the Bar. *CBC/Radio-Canada.* www.cbc.ca/sports/olympics/summer/trackandfield/debbie-brill-high-jump-1.4194545

Simonton, D. K. (2012). Foresight, Insight, Oversight, and Hindsight in Scientific Discovery: How Sighted Were Galileo's Telescopic Sightings? *Psychology of Aesthetics, Creativity, and the Arts, 6,* 243–254.

Simonton, D. K. (2015). Thomas Alva Edison's Creative Career: The Multilayered Trajectory of Trials, Errors, Failures, and Triumphs. *Psychology of Aesthetics, Creativity, and the Arts, 9*, 2–14.

Skinner, B. F. (1966). The Phylogeny and Ontogeny of Behavior. *Science, 153*, 1205–1213.

Skinner, B. F. (1969). *Contingencies of Reinforcement: A Theoretical Analysis*. New York: Appleton-Century-Crofts.

Skinner, B. F. (1974). *About Behaviorism*. New York: Random House.

Stahlman, W. D. and Leising, K. J. (2018). The Coelacanth Still Lives: Bringing Selection Back to the Fore in a Science of Behavior. *American Psychologist, 73*, 918–929.

Thompson, D. (2019, November). Thomas Edison's Greatest Invention. *The Atlantic*.

Verschoth, A. (1971, February 22). She Gets Her Back up. *Sports Illustrated.* https://vault.si.com/vault/1971/02/22/she-gets-her-back-up

Wasserman, E. A. (2012). Species, Tepees, Scotties, and Jockeys: Selected by Consequences. *Journal of the Experimental Analysis of Behavior, 98*, 213–226.

Wasserman, E. A. and Blumberg, M. S. (2010). Designing Minds: How Should We Explain the Origins of Novel Behaviors? *American Scientist, 98*, 183–185.

27 Are We Just Making It Up as We Go Along?

In earlier portions of the book, I dropped a few breadcrumbs. Now, it's time to pick them up and to make a final pitch concerning the accidental and improvisational nature of human experience and innovation. Let's start with the breadcrumbs:

> From the discussion of the Gulf oil spill, recall BP executive Kent Wells' humble confession that *"We will just learn as we go."*
>
> Now, recall Stanley Dudrick's comments on his often-fraught development of Total Parenteral Nutrition, "We're dealing with an evolving medical science. *As we go along, we modify things and learn."*
>
> And, recall Debby Brill's remarks on her development of the Brill Bend, "Back then, sport was not as mainstream as it is today, and *we had to make it up as we went along."*

> Now, it's time for my pitch.

> First, no matter how much we'd like to believe we can anticipate and control our future, life is inherently unpredictable – it throws us curves, often at the most inopportune moments. Just consider the upended future of Jack Westrope (his leg injured in a starting gate accident), Glen Burke (his being unexpectedly traded away from the Dodgers), Virginia Apgar (the 1929 stock market crash crushing her dreams of becoming a surgeon), and Ignacio Ponseti (the Spanish Civil War forcing him to emigrate in order to pursue his medical career). Although we may exert every effort to anticipate and to plan for the future, we must continually amend our actions to contend with unexpected exigencies of immediate importance. Failure to do so might effectively prevent our even being around to enjoy the fruits of any long-term plans.
>
> Second, our current behavior is at least as likely to be affected by its past effects as by its projected results. Indeed, without being able to call on our prior experiences, we would have no basis on which to forecast and prepare for future events.

Third, the intricate interplay between past and future critically leverages adaptive behavioral mechanisms we human beings share with nonhuman animals. The brains of both humans and nonhumans have evolved as indispensable weapons for doing battle with nature's many challenges. Contemporary science continues to shed new light on those mechanisms, bringing into bold relief just how strongly we are bound to obey the natural laws of behavior.

So, are we just making it up as we go along? Indeed, we are! But, this reality requires no apology. We are highly intelligent and flexible beings, not mere flotsam tossed about by choppy waters! We must adapt to a world filled with both perils and opportunities. Our success is not assured; yet, we can improve our chances of success by nimbly adjusting our actions as circumstances warrant. This necessarily trial-and-error process can seem haphazard and awkward; yet, it generally serves us well and should occasion neither laughter nor ridicule, but respect. Critically, it also helps to shape the very behavioral innovations that this book has described and showcased.

Of course, full appreciation and explanation of creative behavior is some distance away. But, if I have made a persuasive case for the involvement of context, consequence, and coincidence, then you'll think twice before invoking "genius" and "foresight" as accounts of the creative behaviors that so captivate us.

Finally, I entitled this last section of the book *Putting it Together*. I did so because of a song that bears the same title. It was written by the Academy Award, Tony Award, and Grammy Award winning composer Stephen Sondheim.

Sondheim's patter-paced lyrics are actually an ode–not to inspiration, but to perspiration. Creating a work of art is no easy matter; doing so demands substantial time, hard work, and persistence. You must begin with a firm foundation. Then, step-by-step, piece-by-piece, you must build upon that foundation. You must continually hone every added bit, so that each one of them signifies a discernible improvement. Sweating all of the countless details in the painstaking process of "putting it together" may ultimately prove to be incommensurate with the final payoff; the 'hit' you seek may turn out to be a 'miss.' But, Sondheim confides, that is truly the state of the art–of making art.

Not only is this progressive and uncertain process the state of the art of making art, it's also the state of life. Embrace it and enjoy the ride! But, please do fasten your seatbelt ... the ride might get a bit bumpy!

Index

Academy of Motion Picture Arts and Sciences, 175
accidental innovation, 52
"acey-deucy" position in horseracing, 49–53, 51f, 294
Adams, Ansel Easton
  advocacy by, 185–187
  artistry of, 188–190
  early influences, 178–180, 293
  evolution as an artist, 190
  fine art vs. practical art, 175–185, 176f
  formative years in photography, 180–185
adaptive action, 7, 297
adaptive behavior, 297–298, 300, 310
Aldrin, Buzz, 199–203, 200f
Alzheimer's Disease, 152
American Academy of Pediatrics, 84
American Broncho-Esophagological Association, 90
American Dental Association, 159
American Film Institute, 281
American Heart Association, 89
American Medical Association, 89
American Red Cross, 89, 91
American Society for Parenteral and Enteral Nutrition (ASPEN), 103
Anamosa State Penitentiary, 273–274
Anesthesia History Association, 67
animal self-medication, 113–119
Apgar score (Apgar, Virginia)
  brief history of Virginia Apgar, 69–75
  consequences of, 67–69
  development of, 65–66, 309
  establishment of, 66–67
Appearance, Pulse, Grimace, Activity, and Respiration (APGAR), 69
Arcaro, Eddie (the "Master"), 50, 52
Armbruster, David A., 31–37, 33f
Armstrong, Neil, 199–203
Ars Gratia Artis motto, 173–174
art for art's sake
  advocacy and, 185–187
  fine art vs. practical art, 174–178

framing of question, 188–190
  photography, 175–185
Aspirin naming process, 106

Baby Jumping festival (El Salto del Colacho), 235–238
Bach, John Sebastian, 248
Baker, Dusty, 55–57
Batten's disease, 125–126
beard growth and shaving. See shaving practices by men
Beauveria bassiana fungus in ants, 115
behavior
  adaptive behavior, 297–298, 300, 310
  creativity and, 3–8, 10, 256, 293, 297
  evolution of, 297–300, 309
  foresight and, 300, 310
  grass-eating behavior by animals, 114–115
  Law of Effect and, 11
  novel behavior, 15
behavioral innovations
  butterfly stroke, 3
  foresight and, 3, 21, 302–303
  Fosbury Flop, 21–26, 288–289
  insight and foresight with, 302–303
  Three C's and, 287
  trial-and-error process with, 310
Bender, Albert M., 183
benzene's chemical structure, 303–306
Bernstein, Leonard, 241
biological evolution in drug naming, 108–110
biostatistical (morphometric) analysis, 253
bipedal locomotion, 205–208
blackout license plates in Iowa
  Dordt University and, 272–274
  natural selection and, 274
  popularity of, 270–271, 271f
Boston Children's Hospital, 125
botch, etymology of, 220–221
Brackin, Francis, 216, 294
brand names of drugs, 106–107

INDEX 313

Brill, Debbie (Brill Bend), 24, 288–290, 293, 309
British Petroleum (BP) oil spill, 8–11, 309
Burke, Glenn, 55–61, 309
Burma-Shave brand, 167–169
Bush, George W., 265
butterfly stroke
    arm stroke, 29
    development of, 30–35, 32–33f, 35f
    dolphin kick, 29–30
    in official events, 36–38

Cammerer, Arno B., 186
Čapek, Josef, 219
Čapek, Karel, 218–220
Carter, Jimmy, 189–190, 262–263, 265
Castellers de Barcelona, 233
castells. *See* human towers
Center for Drug Evaluation and Research (CDER), 107
Centers for Disease Control and Prevention, 89
Chasing and Racing (Cox), 44–46
chemical names of drugs, 106–107
childbed fever, 294
chimpanzee walking studies, 207
Chitwood, Dan, 252–254
Clinton, Hillary, 266, 268
clubfoot treatment, 77–86, 83f, 86f
Cohan, George M., 41
coincidence
    defined, 15
    development of, 293–295
    Fosbury Flop, 21–26, 287–288, 293
    impact of, 149, 282
College Swim Coaches Association of America (CSCAA), 32
Collins, Francis, 121–122, 124–125
Congenital Clubfoot: Fundamentals of Treatment (Ponseti), 84
'consciously visualized' photograph, 182
consequence
    defined, 15
    development of, 295–298
    Fosbury Flop, 287
    impact of, 149, 282
    selection by, 297–298
    unintended consequence, 7, 261–264
consumer demand, 253–254
context
    defined, 15
    development of, 288–292

Fosbury Flop, 287
    impact of, 149, 282
Costner, Kevin, 279–282
Cox, Harding Edward de Fonblanque, 43–47
creative acts by design, 3
creativity
    behavior and, 3–8, 10, 256, 293, 297
    foresight and, 5–8, 14, 16
    genius and, 5–8, 14, 16, 292, 310
    insight and, 5–8
    inspiration and, 5–8, 41
    psychological writings of, 303–304
    scientific understanding of, 302
Crimean War (1853-1856), 138–139, 144, 293
culture
    advances in, 14
    of Catalonia, 232–235
    exotic cultures, 97
    individualism and, 46–47
    subculture, 53

Daniel, Jeffrey, 204–205
Darwin, Charles, 5–6, 10, 108–109, 163–164, 253, 290–291, 295–297
data visualization
    Pasteur, Louis, 133
    Semmelweis, Ignaz, 133–137
Data Visualization Society (DVS), 148–149
Dawkins, Richard, 298, 301–302
de La Borde, Jean Benjamin, 250
Democratic National Committee (DNC), 266–267
Democratic National Convention (1968), 261
Democratic National Convention (1972), 261
Dennett, Daniel, 299–300
dental floss/flossing, 152–159, 157f
design, defined, 300–302
Deutsch, Diana, 246
Dietary Guidelines for Americans (2020), 158
Dietz, Howard, 173–174
Dordt University, 272–274
drug naming process
    Aspirin, 106
    brand names, 106–107
    chemical names, 106–107
    drug costs and, 110
    generation and selection of, 108–110
    generic names, 106–107
    LASA confusions in, 107–108
Dudrick, Stanley J., 97–104, 100f, 309

eating and undernourishment, 97–104
Edison, Thomas, 9, 292
El Salto del Colacho. See Baby Jumping festival
Els Castells. See human towers
Encierro. See running of the bulls
environmentalism, 176–177, 179, 186, 189–190
ethimolegia, defined, 212
etymology and word origins
    botch, 220–221
    ignoramus, 214–217
    malapropism, 212–214
    robot, 217–220
evolution
    of art/artists, 190
    of behavior, 297–300, 309
    biological evolution, 109
    of violin, 250–256, 296, 299–300
experimentation
    butterfly stroke, 34–35
    in dance technique, 193–195, 197
    Fosbury Flop, 289
    Heimlich manuever, 93
    in photographic technique, 180–185
    tooth-cleaning kit, 153
    trial-and-error experimentation, 82, 91, 115–116, 195, 287
Ezingeard, Denis, 108–109

failure, 9, 110, 186, 221, 287, 309
Faril, Juan, 80, 294
Farr, William, 142–143
Fédération Internationale de Natation (FINA), 36
Feste romane (Respighi), 223
Field of Dreams (film), 276, 282
Field of Dreams Movie Site, 281–282
Fiesta del Colaacho: Una farsa castellana (Calvo), 237–238
fine art *vs.* practical art, 174–178
Fiske, George, 180
Food and Drug Administration (FDA), 106–107, 125
food-fight festival. See Tomatina festival
foresight
    acey-deucy and, 53
    of Ansel Adams, 158
    behavior and, 300, 310
    behavioral innovations and, 3, 21, 302–303

creativity and, 5–8, 14, 16
intelligent design and, 296
learning from, 10–11
monkey crouch and, 47
planning with, 197, 260
species generality of, 158
Fosbury, Dick (Fosbury Flop), 21–26, 287–293

Gautier, Théophile, 174–175
gender equality in medical careers, 74
generic names of drugs, 106–107
genetic drift, 253
genius
    behavior and, 310
    creativity and, 5–8, 14, 16, 292, 310
    inspiration and, 292
    Law of Effect and, 5–6
    natural selection and, 290–291
    Ponseti Method and, 85
    puppetry by Basil Twist, 192–197
genomic testing, 121–124
germ theory of disease, 149
gingivitis bacterium (*Porphyromonas gingivalis*), 152
Goldwyn Pictures Corporation, 173–174
gorilla walking studies, 206–207
grass-eating behavior by animals, 114–115

handwashing, as sanitation procedure, 141
Haweiss, Hugh Reginald, 250
Heimlich Maneuver (Heimlich, Henry Judah)
    origin of, 89–94
    procedure, 89, 294–295
    trial-and-error method and, 94–95
Hemingway, Ernest, 223–225
Hemingway, Hadley, 223–225
Herbert, Sidney, 137–142, 149, 293
High Five hand slap, 55–61
Hippocrates of Kos, 77–78, 85, 127–128
homeostasis, defined, 113
Homo neanderthalensis, 118
homophobia, 59–60
homosexuality, 59–60, 241
horseracing
    "acey-deucy" position in, 49–53, 51*f*, 294
    jockey positions in, 39–47, 49–53
    "monkey crouch" position in, 39–47
Houghton Mifflin, 278
Huffman, Michael, 114, 117–118
Human Genome Project, 121–124

human towers (*Els Castells*), 232–235
Hutchings, James Mason, 180, 293

Ickes, Harold L., 185–186
ignoramus, etymology, 214–217, 294
Imhotep (Egyptian polymath), 112–113
imitation, 33
innovation, 3, 52. *See also* behavioral innovations
insight
  acey-deucy and, 52
  behavior and, 6
  case against, 302–303
  copying vs., 118
  creativity and, 3–8
  into origins of TPN, 103
  Ponseti Method and, 85
  problem-solving through, 13
  in scientific discovery, 304, 306
  shaving practices, 165
  trial-and-error vs., 144
  violin construction, 255–256, 296
inspiration
  biological evolution and, 109
  creativity and, 5–8, 41
  genius and, 292
  in scientific development, 304
intelligent design, 295–300
International Anesthesia Research Society, 68
International Nonproprietary Name Program, 106–107
Internet Movie Database, 280
internet role, 84, 112
Iowa Caucus, rise and demise, 259–268
Iowa Democratic Party (IDP), 266–267
ixxéo healthcare, 108–109

Jackson, Joe (Shoeless), 277, 283
Jackson, Michael, 199, 204–205
James I, King, 216–217
jockey positions in horseracing, 39–47, 49–53
Jones, William LaRue, 244
A Journey to Bath (Sheridan), 214
Joyce, James, 9

Kekulé, Friedrich August, 303–306
Kessenich, Larry, 278, 282
Kinsella, W. P. (Bill)
  Field of Dreams, 276, 282

Field of Dreams Movie Site, 281–282
Shoeless Joe Comes to Iowa, 276–278, 282–283
Kolletsschka, Jakob, 294

La Tomatina. *See* Tomatina festival
lamington speed-eating competition, 88–89
Lasorda, Tommy, 59
Lasorda, Tommy, Jr., 59
Law of Effect, 5–6, 9–11, 13–14, 254–255, 290, 295–298
Law of Natural Selection, 5–6, 10, 274, 290–291, 295–297
Little Johnny Jones musical, 41
look-alike, sound-alike (LASA) confusions in drug naming, 107–108
lunar walking, 199–203, 200f, 202f
luthier, 251–254, 299–302

macaque monkeys and dental flossing, 155–158, 157f
Mademoiselle de Maupin (Gautier), 174–175
Makovec, Mila, 125–126
Makris, Nick, 250–254
malapropism, etymology, 212–214
manipulative correction, 78
March of Dimes, 68, 73–74
McGovern, George, 262
McGovern-Fraser Commission, 261
Menand, Louis, 5
mental floss, 152
Merton, Robert, 291, 302–303
mistakes, 9, 210, 268
'monkey crouch' position in horseracing, 39–47
Monolith, the Face of Half Dome (Adams), 182–183
Mooallem, Jon, 60
mood and substance in photography, 183
moonwalking dance move, 204–205
Morcuende, Jose, 82f, 85
The Murders in the Rue Morgue (Poe), 162
music
  photography and, 179–180, 182–183
  Tchaikovsky, Pyotr Ilyich, 243f, 248–249
  violin, evolution of, 250–256, 296, 299–300
Muskie, Edmund, 262
mutations, 251–253, 255, 297

"N-of-one" studies, 126–127
'Name Engineering' model, 108–109
National Aeronautics and Space Administration (NASA), 199–203, 200f, 202f
National Center for Biotechnology Information, 88
National Film Registry, 281
National Institutes of Health, 121–122, 124
natural selection. *See* Law of Natural Selection
Neff, Myles, 50–51
Nightingale, Florence
 happenstance and, 293
 importance of preparation, 149
 important findings of, 137–145
 as "Passionate Statistician," 145–148
 privileged upbringing of, 145–148
"Nightingale rose" diagram, 142–144
Nikisch, Arthur, 244–246
Notes on Matters Affecting the Health, Efficiency, and Hospital Administration of the British Army (Nightingale), 143
novel behavior, 15

Obama, Barack, 10, 123, 265
Olympic Games, 3, 24, 29, 36–37
On the Origin of Species (Darwin), 290
On the Origin of Tepees (Hughes), 290
onomatopoeia, defined, 106
origin. *See also* etymology and word origins
 Heimlich Maneuver, 89–94
 personalized/precision medicine, 127–128
 of puppetry, 192–197
 of Total Parenteral Nutrition, 97–104
origin stories, 21–22, 28, 60
Out at Home (Burke), 59

Parmelian Prints of the High Sierras (Adams), 183, 188
Parmly, Levi Spear, 152–159
Partial Gravity Simulator (POGO), 201, 202f
Pasteur, Louis, 133, 149
Pathétique (Tchaikovsky's Symphony No. 6), 243f, 248–249
Percas-Ponseti, Helena, 80–81
Personalized Medicine Coalition, 123

personalized/precision medicine
 historical origins of, 127–128
 Human Genome Project, 121–124
 promise of, 122–124
 success of, 124–127
pharmacogenomics, 121–122
Phelps, Michael, 3, 28, 37–38
photography art, 175–185. *See also* Adams, Ansel Easton
photorealistic perspective, 177
Poe, Edgar Allen, 162
political party caucuses, 259. *See also* Iowa Caucus
Ponseti International Association at The University of Iowa, 84
Ponseti Vives, Ignacio (Ponseti Method for clubfoot), 77–86, 83f, 86f, 294, 309
Popper, Karl, 299–300
post factum ('after the fact') explanations, 302–303
practical art, 174–178
Practical guide to the management of the teeth: Comprising a discovery of the origin of caries, or decay of the teeth, with its prevention and cure, A (Parmly), 153
Precision Medicine Initiative, 123
Presidential Medal of Freedom, 189–190
primate self-medication, 117–118
puppetry, origins and development, 192–197
Putting it Together (Sondheim), 310–311

quadrupedal locomotion, 207–208
Quande, Bruce, 24–25, 288–289

Rachmaninoff, Sergei Vasilyevich, 247–248
Ramón y Cajal, Santiago, 79
Reagan, Ronald W., 187
representational puppetry, 194
Respighi, Ottorino, 223
retail politicking, 265–269
retention, 47
Richard, James Rodney, 55
The Rivals (Sheridan), 213–214
Robinson, Jackie, 58
Robinson, Phil Alden, 278–280, 282
robot, etymology, 217–220
Rocke, A.J., 304–306
Roosevelt, Franklin D., 186
Ruggle, George, 216–217, 294

ruminant self-medication, 116–117
running of the bulls (*Los Sanfermines Festival*), 223–229
R.U.R. (Čapek), 218–220
Russell, Christopher, 247

Sanders, Bernie, 266
Sanfermines festival, 223–229
scale illusion, 245–246
selection by consequences, 297–298
self-medication
    by animals, 113–119
    Beauveria bassiana fungus in ants, 115
    grass-eating behavior, 114–115
    by *Homo neanderthalensis*, 118
    medical history and prehistory, 112–113
    by primates, 117–118
    by ruminants, 116–117
    zoopharmacognosy, 114
Semmelweis, Ignaz, 133–137, 294
serendipity, 46, 53, 190, 194–197, 223, 225–226, 276, 294
sexual dimorphism in animal kingdom, 162–163
sexual selection, 162–163
shaving practices by men, 162–169, 166f
Shaw, George Bernard, 282
Sheridan, Frances, 214
Sheridan, Richard Brinsley, 213–214
Shoeless Joe Comes to Iowa (Kinsella), 276–278, 282–283
Shoemaker, Bill, 52
Sieg, Jack, 34–37
Sierra Club, 181, 185–186, 188–189
Sierra Nevada: The John Muir Trail (Adams), 186, 189
Simms, Willie, 42–43
Skinner, B.F., 13, 255, 297–298
Sloan, James Forman 'Tod,' 40–47, 293
Smart, Thomas, 215–216, 294
Sondheim, Stephen, 310–311
Spanish festivals
    Baby Jumping festival, 235–238
    human towers, 232–235
    running of the bulls, 223–229
    Tomatina festival, 229–232
Steindler, Arthur, 80, 294
Stieglitz, Alfred, 184
Stradivari, Antonio, 253–254

survival of the fittest, 24–26, 250
Symphonie Fantastique with puppetry, 193, 195–197, 293

Tchaikovsky, Pyotr Ilyich, 243f, 248–249
Thorndike, Edward L., 9–11
Three Cs. *See* coincidence; consequence; context
Tod Sloan: By himself (Sloan), 42
Tomatina festival, 229–232
Total Parenteral Nutrition (TPN), 97–104
town-gown friction, 214–217
trial-and-error learning
    behavioral innovations and, 310
    butterfly stroke, 37–38
    evolution of violin, 255
    Fosbury Flop, 21–26, 287–288
    Heimlich Maneuver, 94–95
    in prehistoric medicine, 113
    Total Parenteral Nutrition, 97–104
    virtual trial-and-error, 299–300
Twist, Basil, 192–197, 293

undernourishment issues, 97–104
Uner Tan Syndrome, 207–208
University of Cambridge, 214–217

variation
    in behavior, 298–300
    in breastroke, 36
    in drug naming, 110
    evolutionary processes of, 43–47, 109, 197, 290–291
    genetic variation, 122
    in Heimlich maneuver, 93
    rose plots and, 142–143
    in ruminant self-medication, 116–117
    in violin design, 251–253, 255
Villalba, Juan, 115–117
violin, evolution of, 250–256, 296, 299–300
Virtual Hospital of The University of Iowa, 84
virtual trial-and-error, 299–300
visualizability, 306
Vuillaume, Jean-Baptiste, 253–254
"vulgar public," 138, 143

walking
    bipedal locomotion and, 205–208
    as dance move, 204–205

walking (cont.)
    development in babies, 208–210
    on lunar surface, 199–203, 200f, 202f
    Uner Tan Syndrome, 207–208
Wallace, Alfred Russell, 290–291
Watt, James G., 187
Weber, Irving B., 31
Webern, Anton, 248
Westrope, Jack, 49–50, 52–53, 294

Westrope, Tommy, 49
Williams, Tennessee, 88
World Health Organization, 106–107

Yosemite Park and Curry Company, 183–184
Yu, Timothy, 125

zeitgeist, defined, 15
zoopharmacognosy, 114